Mathematical Theory of Finite Elements

Computational Science & Engineering

The SIAM series on Computational Science and Engineering publishes research monographs, advanced undergraduate- or graduate-level textbooks, and other volumes of interest to an interdisciplinary CS&E community of computational mathematicians, computer scientists, scientists, and engineers. The series includes introductory volumes aimed at a broad audience of mathematically motivated readers interested in understanding methods and applications within computational science and engineering, monographs reporting on the most recent developments in the field, and volumes addressed to specific groups of professionals whose work relies extensively on computational science and engineering.

SIAM created the CS&E series to support access to the rapid and far-ranging advances in computer modeling and simulation of complex problems in science and engineering, to promote the interdisciplinary culture required to meet these large-scale challenges, and to provide the means to the next generation of computational scientists and engineers.

Mathematical Theory of Finite Elements

LESZEK F. DEMKOWICZ

The University of Texas at Austin
Austin, Texas

Society for Industrial and Applied Mathematics
Philadelphia

Publications Director   Kivmars H. Bowling
Executive Editor   Elizabeth Greenspan
Acquisitions Editor   Elizabeth Greenspan
Developmental Editor   Rose Kolassiba
Managing Editor   Kelly Thomas
Production Editor   Louis R. Primus
Copy Editor   Susan Fleshman
Production Coordinator   Cally A. Shrader
Compositor   Cheryl Hufnagle
Graphic Designer   Doug Smock

Library of Congress Cataloging-in-Publication Data

Names: Demkowicz, Leszek., author.
Title: Mathematical theory of finite elements / Leszek F. Demkowicz, the
    University of Texas at Austin, Austin, Texas.
Description: Philadelphia : Society for Industrial and Applied Mathematics,
    [2023] | Series: Computational science and engineering ; 28 | Includes
    bibliographical references and index. | Summary: "Provides the
    foundations of the mathematical theory for conforming finite element
    methods. Focus is on the concept of discrete stability and the exact
    sequence conforming elements covering both coercive and non-coercive
    problems"-- Provided by publisher.
Identifiers: LCCN 2023025655 (print) | LCCN 2023025656 (ebook) | ISBN
    9781611977721 (paperback) | ISBN 9781611977738 (ebook)
Subjects: LCSH: Finite element method. | AMS: Mathematics education --
    Numerical mathematics -- Numerical analysis.
Classification: LCC QC20.7.F56 D46 2023 (print) | LCC QC20.7.F56 (ebook)
    | DDC 518/.25--dc23/eng/20230817
LC record available at *https://lccn.loc.gov/2023025655*
LC ebook record available at *https://lccn.loc.gov/2023025656*

 is a registered trademark.

Contents

List of Figures

List of Tables

Preface

This monograph is based on my personal lecture notes for the graduate course on *Mathematical Theory of Finite Elements* (EM394H) that I have been teaching at ICES (now the Oden Institute), at the University of Texas at Austin, since 2005. The class has been offered in two versions. The first version is devoted to a study of the energy spaces corresponding to the exact grad-curl-div sequence. The class is rather involved mathematically, and I taught it only every three or four years; see [27] for the corresponding lecture notes. The second, more popular version is covered in the presented notes.

The primal focus of my lectures has been on the concept of *discrete stability* and variational problems set up in the energy spaces forming the exact sequence: H^1-, $H(\text{curl})$-, $H(\text{div})$-, and L^2-spaces. From the application point of view, discussions are wrapped around the classical model problems: diffusion-convection-reaction, elasticity (static and dynamic), linear acoustics, and Maxwell equations. I do not cover transient problems, i.e., all discussed wave propagation problems are formulated in the frequency domain. In the exposition, I follow the historical path and my own personal path of learning the theory. We start with coercive problems for which the discrete stability can be taken for granted, and the convergence analysis reduces to the interpolation error estimation. I cover H^1-, $H(\text{curl})$-, $H(\text{div})$-, and L^2-conforming finite elements and construct commuting interpolation operators.

We then venture into noncoercive problems starting with the fundamental Babuška Theorem and the Mikhlin theory on asymptotic stability. I spend a considerable amount of time on Brezzi's theory for mixed problems and study carefully its relations with the Babuška Theorem.

Finally, I converge to the adventure of my lifetime—the Discontinuous Petrov–Galerkin method co-invented with Jay Gopalakrishnan.

I focus exclusively on conforming methods and a priori error estimation.

The class is taught in a seminar style with the final grade determined by the number of points accumulated for solving the homework problems which essentially complement the lectures. I have always been meeting with students for a weekly discussion session covering the problems and their solutions. I have solved all the homework problems myself securing a methodology consistent with the lectures. If you intend to use the lecture notes for teaching the subject, you may want to ask me for the Solution Manual.

Different parts of these notes have been read by Stefan Henneking, Jaime Mora-Paz, Judit Muñoz, Jiaqi Li, Jacob Salazar, Jacob Badger, and Jonathan Zhang. I am greatly indebted to them for helping to eliminate endless errors and typos and to improve several parts of the manuscript. The notes reflect my own personal growth throughout my career and by no means are intended to compete with a multitude of excellent textbooks on the subject, including the classic books of Ph. Ciarlet [18] and J. T. Oden and J. N. Reddy [67], the fundamental monograph by S. Brenner and L. R. Scott [10], and the fantastic series by A. Ern and J.-L. Guermond [43, 44, 45].

Leszek F. Demkowicz
Austin, August 20, 2023

Chapter 1

Preliminaries

Variational Formulations

This is a very preliminary chapter directed mainly at an engineering audience. We start with a refresher on the classical calculus of variations leading to the concept of a variational (weak) formulation for a boundary-value problem. We quickly descend then on the formalism of the abstract variational formulation in a Hilbert space setting and introduce right away the Galerkin method. We provide two examples of model boundary-value problems—a diffusion-convection-reaction problem and linear elasticity—and derive the corresponding classical variational formulations (Principle of Virtual Work). Finally, in the last section we introduce two more model problems, linear acoustics and Maxwell equations, and revisit elastodynamics, all formulated as systems of first order PDEs. For each of the problems, we introduce then the strong (trivial), mixed, reduced, and ultraweak (UW) variational formulations leading to different energy settings. The last section may be of interest for a more mathematically advanced audience as well.

1.1 ▪ Classical Calculus of Variations

See the book by Gelfand and Fomin [48] for a superb exposition of the subject.

The classical calculus of variations is concerned with the solution of the constrained minimization problem:

$$
\begin{cases}
\text{Find } u(x), x \in [a, b], \text{ such that} \\
u(a) = u_a \,, \\
J(u) = \inf_{w(a)=u_a} J(w) \,,
\end{cases}
\tag{1.1.1}
$$

where the *cost functional* $J(w)$ is given by

$$
J(w) = \int_a^b F(x, w(x), w'(x)) \, dx \,.
\tag{1.1.2}
$$

Integrand $F(x, u, u')$ may represent an arbitrary scalar-valued function of three arguments:[1] x, u, u'. Boundary condition (BC), $u(a) = u_a$, with u_a given, is known as the *essential BC*.

---

[1]Note that, in this classical notation, x, u, u' stand for the arguments of the integrand. We could have used any other three symbols, e.g., x, y, z.

In the following discussion we sweep all regularity considerations under the carpet. In other words, we assume whatever is necessary to make sense of the considered integrals and derivatives.

Assume now that $u(x)$ is a solution to problem (1.1.1). Let $v(x), x \in [a, b]$, be an arbitrary *test function*. Function

$$w(x) = u(x) + \epsilon v(x)$$

satisfies the essential BC if and only if (iff) $v(a) = 0$, i.e., the test function must satisfy the *homogeneous essential BC*. Consider an auxiliary function,

$$f(\epsilon) := J(u + \epsilon v).$$

If functional $J(w)$ attains a minimum at u, then function $f(\epsilon)$ must attain a minimum at $\epsilon = 0$ and, consequently,

$$\frac{df}{d\epsilon}(0) = 0.$$

It remains to compute the derivative of function

$$f(\epsilon) = J(u + \epsilon v) = \int_a^b F(x, u(x) + \epsilon v(x), u'(x) + \epsilon v'(x)) \, dx.$$

By the Leibniz formula (see, e.g., [51, p. 17]),

$$\frac{df}{d\epsilon}(\epsilon) = \int_a^b \frac{d}{d\epsilon} F(x, u(x) + \epsilon v(x), u'(x) + \epsilon v'(x)) \, dx,$$

so, utilizing the chain formula, we get

$$\frac{df}{d\epsilon}(\epsilon) = \int_a^b \left\{ \frac{\partial F}{\partial u}(x, u(x) + \epsilon v(x), u'(x) + \epsilon v'(x)) v(x) \right. $$
$$\left. + \frac{\partial F}{\partial u'}(x, u(x) + \epsilon v(x), u'(x) + \epsilon v'(x)) v'(x) \right\} \, dx.$$

Setting $\epsilon = 0$, we get

$$\frac{df}{d\epsilon}(0) = \int_a^b \left\{ \frac{\partial F}{\partial u}(x, u(x), u'(x)) v(x) + \frac{\partial F}{\partial u'}(x, u(x), u'(x)) v'(x) \right\} \, dx. \qquad (1.1.3)$$

Again, remember that u, u' in $\partial F/\partial u, \partial F/\partial u'$ denote simply the derivatives of integrand F with respect to the second and third arguments of F. Derivative (1.1.3) is identified as the *directional derivative* of functional $J(w)$ in the direction of test function $v(x)$. The linear operator,

$$v \to \langle (\partial J)(u), v \rangle := \frac{df}{d\epsilon}(0) = \int_a^b \left(\frac{\partial F}{\partial u}(u(x), u'(x)) v(x) + \frac{\partial F}{\partial u'}(u(x), u'(x)) v'(x) \right) \, dx,$$
$$(1.1.4)$$

is identified as the *Gâteaux differential* of $J(w)$ at u.

The necessary condition for u to be a minimizer now reads as follows:

$$\begin{cases} u(a) = u_a \\ \langle (\partial J)(u), v \rangle = \int_a^b \left(\frac{\partial F}{\partial u}(x, u, u') v + \frac{\partial F}{\partial u'}(x, u, u') v' \right) \, dx = 0 \quad \forall v : v(a) = 0. \end{cases} \qquad (1.1.5)$$

Integral identity (1.1.5), which has to be satisfied for any eligible test function v, is identified as the *variational formulation* corresponding to the minimization problem.

It turns out that the variational formulation is equivalent to the corresponding *Euler–Lagrange (E-L)* differential equation and an additional *natural BC* at $x = b$. The key tool to deriving both of them is the following Fourier's Lemma.

Lemma 1.1 (Fourier's Lemma). *Let $f \in C[a, b]$ such that*

$$\int_a^b f(x)v(x)\, dx = 0$$

for every test function $v \in C[a, b]$ vanishing at the endpoints: $v(a) = v(b) = 0$.
Then $f(x) = 0$, $x \in [a, b]$.

Proof. See [66, p. 531].   □

In order to apply Fourier's argument, we need first to move the derivative from the test function in the second term in (1.1.5). We get

$$\int_a^b \left(\frac{\partial F}{\partial u}(x, u, u') - \frac{d}{dx}\frac{\partial F}{\partial u'}(x, u, u') \right) v\, dx + \frac{\partial F}{\partial u'}(x, u(x), u'(x))v(x)|_a^b = 0.$$

But $v(a) = 0$ so the boundary terms reduce only to the term at $x = b$ (we do not test at $x = a$),

$$\int_a^b \left(\frac{\partial F}{\partial u}(x, u, u') - \frac{d}{dx}\frac{\partial F}{\partial u'}(x, u, u') \right) v\, dx + \frac{\partial F}{\partial u'}(b, u(b), u'(b))v(b) = 0. \qquad (1.1.6)$$

We can follow now with the Fourier argument.
Step 1: Assume additionally that we test only with test functions that vanish at *both* $x = a$ and $x = b$. The boundary term in (1.1.6) disappears and, by Fourier's Lemma, we can conclude that

$$\frac{\partial F}{\partial u}(x, u(x), u'(x)) - \frac{d}{dx}\frac{\partial F}{\partial u'}(x, u(x), u'(x)) = 0. \qquad (1.1.7)$$

We say that *we have recovered the differential equation.*
Step 2: Once we know that the function above vanishes, the integral term in (1.1.6) must vanish *for any* test function v. Consequently

$$\frac{\partial F}{\partial u'}(b, u(b), u'(b))v(b) = 0$$

for any v. Choose such a test function that $v(b) = 1$ to learn that the solution must satisfy the *natural BC* at $x = b$,

$$\frac{\partial F}{\partial u'}(b, u(b), u'(b)) = 0. \qquad (1.1.8)$$

We have recovered the natural BC. The E-L equation (1.1.7) along with the essential and natural BCs constitute the *Euler–Lagrange Boundary-Value Problem* (E-L BVP),

$$\begin{cases} \qquad\qquad\qquad\qquad u(a) = u_a & \text{(essential BC)}, \\ \dfrac{\partial F}{\partial u}(x, u, u') - \dfrac{d}{dx}\left(\dfrac{\partial F}{\partial u'}(x, u, u') \right) = 0 & \text{(E-L equation)}, \\ \qquad\qquad \dfrac{\partial F}{\partial u'}(b, u(b), u'(b)) = 0 & \text{(natural BC)}. \end{cases} \qquad (1.1.9)$$

Neglecting the regularity issues, we can say that the E-L BVP and variational formulations are in fact equivalent to each other. Indeed, we have already shown that the variational formulation implies the E-L BVP. To show the converse, we multiply the E-L equation with a test function $v(x)$, integrate it over interval (a, b), and add to it the natural BC multiplied by $v(b)$. We then integrate (back) by parts, to arrive at the variational formulation. We say that the variational formulation and the E-L BVP are *formally equivalent*, formally meaning without paying attention to regularity assumptions.

The E-L BVP provides a foundation for finite difference discretizations, whereas the variational formulation is a starting point for the Galerkin method and finite elements.

Exercises

1.1.1. Consider a slight variation of Fourier's Lemma.

Lemma 1.2. *Let $f \in C[a, b]$ such that*

$$\int_a^b f(x)v(x)\,dx = 0$$

for every test function $v \in C[a, b]$. Then $f(x) = 0$, $x \in [a, b]$.

Which of the two lemmas, Lemma 1.1 or the lemma above, is *stronger*? Prove the lemma above (one line argument). (1 point)

1.1.2. Derive the variational formulation and the corresponding E-L BVP for the minimization problem:

$$\begin{cases} u(a) = u_a, u'(a) = d_a\,, \\ J(u) := \displaystyle\int_a^b F(x, u, u', u'')\,dx \to \min\,. \end{cases}$$

Discuss other possible essential BCs. *Hint:* In this and the next problems, you will need a more general version of Fourier's Lemma.

Lemma 1.3 (Fourier's Lemma Generalized). *Let $\Omega \subset \mathbb{R}^N$ be a Lipschitz domain. Let $f \in L^2(\Omega)$ be such that*

$$\int_\Omega fv = 0 \quad \forall v \in C_0^\infty(\Omega)\,,$$

where $C_0^\infty(\Omega)$ denotes the space of all C^∞ functions with compact support in Ω; see [27]. Then

$$f = 0 \quad \text{almost everywhere (a.e.) in } \Omega\,.$$

If $f(x)$ is continuous then vanishing a.e. implies that $f(x) = 0$ in Ω.

Proof. The result is a straightforward consequence of density of $C_0^\infty(\Omega)$ in $L^2(\Omega)$ [27]. □

(3 points)

1.1.3. Derive the variational formulation and the corresponding E-L BVP for the two-dimensional (2D) minimization problem:

$$\begin{cases} u = u_0 \text{ on } \Gamma_1 \,, \\ \int_\Omega F\left(x, y, u(x,y), \dfrac{\partial u}{\partial x}(x,y), \dfrac{\partial u}{\partial y}(x,y)\right) \, dxdy \to \min \,. \end{cases}$$

Here $\Omega \subset \mathbb{R}^2$ is a bounded 2D domain with boundary Γ split into two disjoint parts, $\Gamma = \Gamma_1 \cup \Gamma_2$. (3 points)

1.1.4. An interface problem. Consider the elastic beam depicted in Fig. 1.1. Deflection $w(x)$ of the beam minimizes the *total potential energy* given by the functional

$$J(w) = \frac{1}{2}\int_0^{3l/2} EI(w'')^2 - \left[\int_0^{3l/2} qw + P_0 w\left(\frac{3l}{2}\right) + M_0 w'\left(\frac{3l}{2}\right)\right]$$

among all possible displacements that satisfy the *kinematic BC*:

$$w(0) = w'(0) = w(l) = 0\,.$$

(i) Derive the Gâteaux derivative of cost functional $J(w)$ and the corresponding variational formulation for the problem.

(ii) Use integration by parts (twice) and the Fourier's Lemma argument to derive the corresponding E-L equation(s) in subintervals $(0, l)$ and $(l, 3l/2)$, BCs at $x = 3l/2$, and interface conditions at $x = l$.

(iii) Show the (formal) equivalence between the variational formulation and the E-L interface boundary-value problem.

(3 points)

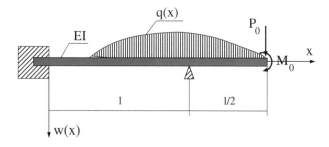

Figure 1.1: An elastic beam example.

1.2 ▪ Abstract Variational Formulation

We begin our study on Galerkin and finite element (FE) methods with the *abstract variational formulation*.

Abstract variational formulation.   Abstract variational formulation reads as follows:

$$\begin{cases} u \in U \,, \\ b(u, v) = l(v) \quad \forall v \in V \,. \end{cases} \tag{1.2.10}$$

Here U is a *trial space*, and V is a *test space*. In this monograph, we shall restrict ourselves to Hilbert spaces only. The two spaces come with inner products and the corresponding (Euclidean) norms,
$$\|u\|_U^2 = (u, u)_U, \qquad \|v\|_V^2 = (v, v)_V.$$
On the left we have a bilinear (or sesquilinear) form $b : U \times V \to \mathbb{R}(\mathbb{C})$ defining the operator, and on the right we have a linear (antilinear) form $l : V \to \mathbb{R}(\mathbb{C})$ specifying the load. It goes without saying that both forms must be continuous. It is easy to show (see Exercise 1.2.1 and Exercise 1.2.2) that the continuity of forms b and l is equivalent to their boundedness, i.e.,

$$|b(u, v)| \leq M\|u\|_U \|v\|_V \qquad \forall u \in U, v \in V \tag{1.2.11}$$

and

$$|l(v)| \leq C\|v\|_V \qquad \forall v \in V \tag{1.2.12}$$

for some $M, C > 0$.

Make sure that, for each variational formulation discussed in the next section, you are able to specify energy spaces U, V and the forms $b(u, v), l(v)$.

Accounting for nonhomogeneous BCs.   In the case of nonhomogeneous essential BCs, we may have to consider a more general abstract variational problem:

$$\begin{cases} u \in \tilde{u}_0 + U, \\ b(u, v) = l(v) \quad \forall v \in V. \end{cases} \tag{1.2.13}$$

Here U is a subspace of a larger energy space X, and $\tilde{u}_0$ is an element of X. The symbol $\tilde{u}_0 + U$ denotes the algebraic sum of $\tilde{u}_0$ and U, known also as an *affine subspace* or *affine submanifold* of X,
$$\tilde{u}_0 + U := \{\tilde{u}_0 + w : w \in U\}.$$
In practice the nonhomogeneous boundary data u_0 is known only on the boundary of the domain. The tilde over u_0 denotes a *finite energy lift* of u_0, i.e., an extension of u_0 to the whole domain that lives in the energy space X. In this discussion though, $\tilde{u}_0$ is simply an arbitrary element of X that does not[2] live in U. The moral of this abstract notation is that solution u can be sought in the form $u = \tilde{u}_0 + w$ where $w \in U$. Substituting this representation of u into the variational formulation, using linearity of form b with respect to the first argument, and moving known terms to the right-hand side, we obtain

$$\begin{cases} w \in U, \\ b(w, v) = \underbrace{l(v) - b(\tilde{u}_0, v)}_{=:l^{\mathrm{mod}}(v)} \quad \forall v \in V. \end{cases}$$

The case of nonhomogeneous BCs can thus be studied within the framework of original formulation (1.2.10) provided we replace the linear form $l(v)$ with the *modified linear form* $l^{\mathrm{mod}}(v)$. This explains also why the essential BC data u_0 is classified as part of the load.

Operator form of the variational formulation.   With every continuous sesquilinear form $b(u, v)$, $u \in U, v \in V$, we can associate a continuous linear operator from trial space U into the *dual* of test space V,

$$B : U \to V', \qquad \langle Bu, v \rangle := b(u, v) \quad u \in U, v \in V. \tag{1.2.14}$$

---

[2]Otherwise, $\tilde{u}_0 + U = U$.

Note the following.

- Operator B is well-defined, i.e., Bu is an element of topological dual V'. Indeed, it represents an antilinear and continuous functional.

- Operator B is linear and continuous. Its norm is equal to the (smallest) continuity constant M for the sesquilinear form.

The abstract variational problem can thus be reformulated as the operator equation,

$$\langle Bu, v \rangle = \langle l, v \rangle, \quad v \in V,$$

or, using the argumentless notation,

$$Bu = l.$$

We can argue that the variational problem is just a special linear operator equation where the operator takes values in a dual space. This observation will provide later the fundamental link between the Babuška–Nečas Theorem and the Banach Closed Range Theorem.

Galerkin approximation. It is not too early to introduce the fundamental concept of the Galerkin approximation of the abstract variational problem. We approximate solution u and test functions v with finite linear combinations:

$$u \approx u_h := \sum_{j=1}^{N} u_j e_j, \qquad v \approx v_h := \sum_{i=1}^{N} v_i g_i, \tag{1.2.15}$$

where *trial basis functions* $e_j \in U$ live in the trial space, the *test basis functions* $g_i \in V$ live in the test space, coefficients $u_j \in \mathbb{R}(\mathbb{C})$ are the unknown *degrees of freedom (d.o.f.)* to be determined, and coefficients v_i are arbitrary real (complex) numbers. Notice that we use the same number of terms in both approximating combinations (explain why?). Symbol h here is a general, abstract discretization symbol. In the context of finite elements, it may be interpreted as mesh size. We simply replace now u with u_h and v with v_h, and request the resulting system to be satisfied for any test function coefficients v_i. We end up with the following system of linear algebraic equations:

$$\sum_{j=1}^{N} \underbrace{b(e_j, g_i)}_{=: b_{ij}} u_j = \underbrace{l(g_i)}_{=: l_i}, \quad i = 1, \ldots, N. \tag{1.2.16}$$

Vector l_i and matrix b_{ij} are known as *load vector* and *stiffness matrix*. The Galerkin method can now be summarized in three steps.

1. Select trial and test basis functions, and compute the stiffness matrix and load vector.

2. Solve the resulting system of linear equations.

3. Compute the approximate solution (1.2.15) using the (now) known d.o.f. and postprocess it as necessary.

The collection of all u_h and v_h of form (1.2.15), for arbitrary d.o.f. u_j, v_i, is identified as the finite-dimensional trial space $U_h \subset U$ and test space $V_h \subset V$. The approximate problem can be written thus in the more concise form

$$\begin{cases} u_h \in U_h, \\ b(u_h, v_h) = l(v_h) \quad \forall v_h \in V_h. \end{cases} \tag{1.2.17}$$

The difference $e_h := u - u_h$ is identified as the *Galerkin error*. The main purpose of this monograph is to study the evolution (convergence) of the Galerkin error

$$\|u - u_h\|_U \to 0 \quad \text{as} \quad h \to 0 \,.$$

Stability of discretization. We shall say that the Galerkin method is *stable* if there exists a *stability constant* $C > 0$ such that

$$\|u - u_h\|_U \le C \underbrace{\inf_{w_h \in U_h} \|u - w_h\|_U}_{=: \text{best approximation error}} \,.$$

If the method is stable, and the best approximation error converges to zero, then so does the error and it converges with the same rate as the best approximation error. We say then also that the *discretization is optimal*. Note that C need not be independent of h. If it blows up with h, the best approximation error should converge faster to zero than $C_h \to \infty$ in order to see the (nonoptimal) convergence.

Try to remember the phrase,

Approximability and stability imply convergence.

Remark 1.4. We differentiate between the case of a symmetric variational setting, i.e., when trial and test space are equal, $U = V$, and the case of an unsymmetric variational setting, i.e., when $U \ne V$. In the first case we talk about a *Bubnov–Galerkin* method and in the second case about a *Petrov–Galerkin* method. Typically, by a (plain) Galerkin method we mean only the Bubnov–Galerkin case.

Exercises

1.2.1. Equivalence of continuity and boundedness for linear (antilinear) forms. Let V be a normed vector space and l be a linear (antilinear) functional defined on V. Prove that the following conditions are equivalent to each other.

   (i) l is continuous on V,

   (ii) l is sequentially continuous on V,

   (iii) l is continuous at 0 (zero vector),

   (iv) l is sequentially continuous at 0,

   (v) l is *bounded*, i.e., there exists $C > 0$ such that

   $$|l(v)| \le C\|v\|_V \,,$$

   where $\|v\|_V$ is the norm in V.

   (3 points)

1.2.2. Equivalence of continuity and boundedness for bilinear (sesquilinear) forms. Let U, V be normed vector spaces and b be a bilinear (sesquilinear) functional defined on $U \times V$. Prove that the following conditions are equivalent to each other.

   (i) b is continuous on $U \times V$,

   (ii) b is sequentially continuous on $U \times V$,

(iii) b is continuous at $(0,0)$,

(iv) b is sequentially continuous at $(0,0)$,

(v) b is *bounded*, i.e., there exists $C > 0$ such that

$$|b(u,v)| \leq C\|u\|_U \|v\|_V .$$

(3 points)

1.2.3. Dual norm. Let V be a normed vector space and l be a continuous (bounded) linear (antilinear) functional defined on V. Let $\|l\|$ be the "smallest" constant that we can use in the boundedness condition,

$$\|l\| := \inf\{C \, : \, |l(v)| \leq C\|v\|_V\} .$$

(i) Prove equivalent characterizations for $\|l\|$,

$$\|l\| = \sup_{v \neq 0} \frac{|l(v)|}{\|v\|} = \sup_{\|v\|=1} |l(v)| = \sup_{\|v\|\leq 1} |l(v)| .$$

(ii) Let V' be the collection of all bounded linear (antilinear) functionals defined on V. Argue that V' is closed with respect to the standard operations on functions and, therefore, constitutes a subspace of algebraic dual V^* consisting of all linear (antilinear) functionals on V. Prove that $\|l\|$ satisfies the axioms for a norm, i.e., V' is a normed space (called the *topological dual* of normed space V).

(3 points)

1.2.4. Let V be a Hilbert space. Prove that the infimum and the suprema in Exercise 1.2.3 are actually attained, i.e., the inf and sup symbols can be replaced with min and max. (3 points)

1.2.5. Space of bounded bilinear functionals. Generalize the concept of the norm of a linear functional to bilinear (sesquilinear) functionals. Let U, V be normed vector spaces and b be a continuous (bounded) bilinear (sesquilinear) functional defined on $U \times V$. Let $\|b\|$ denote the "smallest" constant that we can use in the boundedness condition,

$$\|b\| := \inf\{C \, : \, |b(u,v)| \leq C\|u\|_U \|v\|_V\} .$$

(i) Prove equivalent characterizations for $\|b\|$,

$$\|b\| = \sup_{u,v \neq 0} \frac{|b(u,v)|}{\|u\| \|v\|} = \sup_{\|u\|=1, \|v\|=1} |b(u,v)| = \sup_{\|u\|\leq 1, \|v\|\leq 1} |b(u,v)| .$$

(ii) Prove that the collection of all bounded bilinear (sesquilinear) functionals defined on $U \times V$ forms a normed space with norm $\|b\|$.

(iii) Let $B : U \to V'$ be the operator corresponding to form $b(u,v)$,

$$\langle Bu, v \rangle := b(u,v) \quad u \in U, \, v \in V .$$

Prove that $\|B\| = \|b\|$.

(iv) Show that the infimum and all the suprema above are attained if U, V are Hilbert spaces.

(3 points)

1.3 ▪ Classical Variational Formulations

1.3.1 ▪ Diffusion-Convection-Reaction Problem

Let $\Omega \in \mathbb{R}^N$, $N = 1, 2, 3$, be a bounded domain (:= open, connected set). Let boundary $\Gamma = \partial\Omega$ be split into two disjoint parts Γ_1, Γ_2. More precisely, Γ_1, Γ_2 are assumed to be (relatively) open in Γ and

$$\Gamma = \overline{\Gamma}_1 \cup \overline{\Gamma}_2, \quad \Gamma_1 \cap \Gamma_2 = \emptyset,$$

where the overbar denotes the closure in Γ.

Consider a general *diffusion-convection-reaction* boundary-value problem,

$$\begin{cases} \text{Find } u = u(x),\ x \in \overline{\Omega},\ \text{such that} \\ -(a_{ij}u_{,j})_{,i} + b_j u_{,j} + cu = f \quad \text{in } \Omega, \\ \hspace{3.5cm} u = u_0 \quad \text{on } \Gamma_1, \\ \hspace{2.2cm} a_{ij}u_{,j}n_i = g \quad \text{on } \Gamma_2. \end{cases} \tag{1.3.18}$$

Here $a_{ij}(x)$, $b_j(x)$, $c(x)$, $x \in \overline{\Omega}$ are the diffusion, convection, and reaction coefficients (*material data*), and functions $f(x), x \in \Omega$, $u_0(x), x \in \Gamma_1$, $g(x), x \in \Gamma_2$ are the *load data*, all assumed to be given. We are using the Einstein summation convention.

Elementary integration by parts formula. The following formula generalizes the classical 1D integration by parts to multispace dimension, and it is the workhorse *for deriving all* variational formulations:

$$\int_\Omega \frac{\partial u}{\partial x_i} v = -\int_\Omega u \frac{\partial v}{\partial x_i} + \int_{\partial\Omega} uvn_i, \tag{1.3.19}$$

where $\Omega \subset \mathbb{R}^N$, $N = 2, 3$, and n_i is the ith component of the outward normal unit vector n. For $N = 2$, the domain integral is a double integral, and the boundary integral is the *line integral of the first type*. For $N = 3$, we are dealing with a triple integral and the *surface integral of the first type*. The line and surface integrals, and the formula, can be generalized to any N dimension.

The elementary integration by parts formula can be used to derive more complicated integration by parts formulas for different differential operators. The most classical ones involve operators of gradient, curl, and divergence.

$$\int_\Omega \operatorname{div} u\, q = -\int_\Omega u\, \nabla q + \int_\Gamma u_n\, q,$$

where $u_n := u_i n_i$ denotes the normal component of vector u. Similarly,

$$\int_\Omega \nabla \times E\, F = \int_\Omega E\, \nabla \times F + \int_\Gamma n \times E\, F.$$

Note that we do not use boldface for vectors (and tensors), and you have to deduce from context what type of functions we are dealing with, and whether we mean product of two numbers, scalar product of two vectors, or contraction of two tensors. Talking about tensors, we have the formula

$$\int_\Omega \operatorname{div} \sigma\, v = -\int_\Omega \sigma\, \nabla v + \int_\Gamma \sigma n\, v.$$

If σ is the stress tensor, then $t := \sigma n$ is the traction vector.

Classical variational formulation. We take an arbitrary test function $v = v(x)$, $x \in \overline{\Omega}$, multiply PDE $(1.3.18)_1$ with $v(x)$, integrate over Ω, and integrate the first term by parts using the elementary integration by parts formula to obtain

$$\int_\Omega a_{ij} u_{,j} v_{,i} + b_j u_{,j} v + cuv - \int_\Gamma a_{ij} u_{,j} n_i v = \int_\Omega fv \,.$$

We can now split the boundary integral into two parts corresponding to Γ_1 and Γ_2. On Γ_2 the *co-normal derivative* $a_{ij} u_{,j} n_i$ is known and we can replace it with the given load data g. On Γ_1, the derivative is unknown a priori, and we eliminate this part of the boundary integral by assuming $v = 0$ on Γ_1. We simply *do not test* on Γ_1. This is also consistent with the concept of *essential BC* in the classical calculus of variations: test functions always satisfy the homogeneous version of the essential BC.

Contrary to the BC on Γ_2 which has been *built into* the formulation, the first BC has to be simply restated. The classical formulation now reads as follows:

$$\begin{cases} \text{Find } u = u(x), \ x \in \overline{\Omega}, \text{ such that} \\[4pt] u = u_0 \quad \text{on } \Gamma_1 \,, \\[4pt] \int_\Omega a_{ij} u_{,j} v_{,i} + b_j u_{,j} v + cuv = \int_\Omega fv + \int_{\Gamma_2} gv \\[4pt] \qquad\qquad \forall\, v \text{ such that } v = 0 \text{ on } \Gamma_1 \,. \end{cases} \qquad (1.3.20)$$

Regularity assumptions. We now have to start paying attention to making appropriate assumptions to guarantee that all terms in the variational formulation are well-defined. The first critical tool is the *Cauchy–Schwarz* inequality,

$$\left| \int_\Omega uv \right| \leq \left(\int_\Omega |u|^2 \right)^{\frac{1}{2}} \left(\int_\Omega |v|^2 \right)^{\frac{1}{2}} , \qquad (1.3.21)$$

where

$$\|u\| := \left(\int_\Omega |u|^2 \right)^{\frac{1}{2}}$$

is identified as the L^2-norm of function u. The L^2-space will be denoted by $L^2(\Omega)$ and the symbol for the space will be omitted in the symbol for the norm, i.e.,

$$\|u\| = \|u\|_{L^2(\Omega)} \,.$$

Recall that the L^2-space is a Hilbert space with the inner product,

$$(u, v)_{L^2(\Omega)} := \int_\Omega u\overline{v}, \quad \|u\|^2 = (u, u) \,.$$

In the discussed case, all functions are real-valued, so the complex conjugate over function v is redundant. We shall skip the space symbol in the inner product notation as well.

If we assume now that the reaction coefficient is bounded,

$$|c(x)| \leq c_{\max} < \infty, \quad x \in \Omega \,,$$

and functions $u, v \in L^2(\Omega)$, the Cauchy–Schwarz inequality implies that the integral corresponding to the reaction term is bounded as well. Indeed,

$$\left| \int_\Omega c(x) uv \right| \leq \int_\Omega |c(x)| \, |u| \, |v| \leq c_{\max} \int_\Omega |u| \, |v| \leq c_{\max} \|u\| \, \|v\| \,.$$

By the same argument, if we assume that diffusion matrix a_{ij} and the advection vector b_j are bounded,

$$\|a(x)\| \le a_{\max}, \quad \|b(x)\| \le b_{\max},$$

we can bound the first two terms on the left-hand side as well,

$$\left| \int_\Omega a_{ij} u_{,i} v_{,j} \right| \le a_{\max} \left(\sum_i \|u_{,i}\|^2 \right)^{1/2} \left(\sum_j \|v_{,j}\|^2 \right)^{1/2},$$

$$\left| \int_\Omega b_j u_{,j} v \right| \le b_{\max} \left(\sum_i \|u_{,i}\|^2 \right)^{1/2} \|v\|.$$

Notice that by $\|b\|$ we mean the norm of the vector,

$$\|b\| - \left(\sum_i |b_i|^2 \right)^{1/2},$$

and by $\|a\|$ the norm of a matrix. Typically, we assume that the diffusion matrix is symmetric. The norm of a is then

$$\|a\| = \max_j |\lambda_j|,$$

where λ_j are the (real) eigenvalues of a. If a is not assumed to be symmetric, then the norm of a is equal to the maximum singular value of a.

These considerations lead to the introduction of our first energy space—the Sobolev space of the first order,

$$H^1(\Omega) := \{ u \in L^2(\Omega) \ : \ \boldsymbol{\nabla} u \in L^2(\Omega) \}. \tag{1.3.22}$$

This is a Hilbert space with inner product

$$(u, v)_{H^1(\Omega)} = (u, v) + \sum_i (u_{,i}, v_{,i})$$

and the norm

$$\|u\|_{H^1(\Omega)}^2 := \|u\|^2 + \sum_i \|u_{,i}\|^2.$$

Summing up, we can claim the estimate:

$$\left| \int_\Omega a_{ij} u_{,j} v_{,i} + b_j u_{,j} v + cuv \right| \le (a_{\max} + b_{\max} + c_{\max}) \|u\|_{H^1(\Omega)} \|v\|_{H^1(\Omega)}. \tag{1.3.23}$$

Proceeding along similar lines, we can also estimate the right-hand side,

$$\left| \int_\Omega fv + \int_{\Gamma_2} gv \right| \le \|f\| \, \|v\| + \|g\|_{L^2(\Gamma_2)} \|v\|_{L^2(\Gamma_2)}$$

with the implicit assumption that $\|f\|$, $\|g\|_{L^2(\Gamma_2)}$ are bounded. It follows from the famous *Trace Theorem* [27] that there exists a positive constant $C > 0$ such that

$$\|v\|_{L^2(\Gamma_2)} \le C \|v\|_{H^1(\Omega)}.$$

This leads to our final estimate of the right-hand side,

$$\left| \int_\Omega fv + \int_{\Gamma_2} gv \right| \le (\|f\| + C \|g\|_{L^2(\Gamma_2)}) \|v\|_{H^1(\Omega)}. \tag{1.3.24}$$

1.3.2 ▪ Linear Elasticity

The *linear elasticity* or, more precisely, the *elastostatics* problem deals with the deformation of an elastic body occupying domain $\Omega \subset \mathbb{R}^N$, $N = 2, 3$, under the load of body forces $f = \{f_i\}$ and tractions $g = \{g_i\}$; see Fig. 1.2.

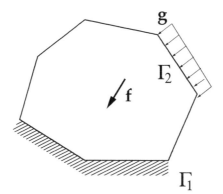

Figure 1.2: An elastic body under the load of volume (body) forces f and surface forces (tractions) g.

The *unknowns* include displacement u_i, strains ϵ_{ij}, and stresses σ_{ij}, $i, j = 1, \ldots, N$. The following equations need to be satisfied.

- Strain-displacement relations:

$$\epsilon_{ij} = \frac{1}{2}(u_{i,j} + u_{j,i}).$$

- Equilibrium (conservation of linear momentum) equations:

$$-\sigma_{ij,j} = f_i.$$

- Conservation of angular momentum:

$$\sigma_{ij} = \sigma_{ji}.$$

- Constitutive equations:

$$\sigma_{ij} = E_{ijkl}\epsilon_{kl} \quad \text{or} \quad \epsilon_{ij} = C_{ijkl}\sigma_{kl},$$

where elasticities satisfy the conditions

$$E_{ijkl} = E_{jikl} = E_{ijlk} \qquad \text{(minor symmetries)},$$
$$E_{ijkl} = E_{klij} \qquad \text{(major symmetry)},$$
$$E_{ijkl}\xi_{ij}\xi_{kl} > 0 \quad \forall \xi_{ij} = \xi_{ji} \neq 0 \quad \text{(positive definiteness)}.$$

- Cauchy stress vector–stress tensor relation:

$$t_i = \sigma_{ij}n_j.$$

We shall consider standard BCs:

- Displacement (kinematic) BC:

$$u_i = 0 \qquad \text{on } \Gamma_1.$$

- Traction BC:

$$t_i = \sigma_{ij} n_j = g_i \qquad \text{on } \Gamma_2.$$

For simplicity, we assume the homogeneous kinematic BCs.

Lamé equations.   Using the strain-displacement relations to represent strains ϵ_{ij} in terms of displacements u_i, and, in turn, Cauchy relations to represent stresses in terms of displacements, we can reduce the whole system to just N differential equations of second order,

$$-(E_{ijkl} u_{k,l})_{,j} = f_i. \tag{1.3.25}$$

The Lamé equations are accompanied with the BCs above.

Classical variational formulation: Principle of Virtual Work.   The classical variational formulation, known in mechanics as the *Principle of Virtual Work*, is derived in a way fully analogous to the one for the diffusion-convection-reaction problem. We multiply equations (1.3.25) with test functions v_i that vanish on Γ_1, integrate over Ω, and integrate by parts. The boundary term reduces to the integral over Γ_2. We build in the traction BC and move the term to the right-hand side. The final formulation looks as follows:

$$\begin{cases} u_i \in H^1(\Omega), u_i = 0 \text{ on } \Gamma_1, \\ \displaystyle\int_\Omega E_{ijkl} u_{k,l} v_{i,j} = \int_\Omega f_i v_i + \int_{\Gamma_2} g_i v_i \qquad v_i \in H^1(\Omega) : v_i = 0 \text{ on } \Gamma_1. \end{cases}$$

Exercises

1.3.1. (Calculus II refresher) Define line and surface integrals of the first type. Discuss why they are identified as *geometrical quantities*. Consider a unit circle with center at origin and density $\rho(x) = |x_2|$. Compute the mass of the circle. Similarly, consider a unit sphere centered at the origin, with density $\rho(x) = |x_3|$. Compute its mass. (2 points)

1.3.2. Use whatever source you need to prove the elementary integration by parts formula (1.3.19) in both two and three space dimensions. (2 points)

1.3.3. Follow the discussion for the diffusion-convection-reaction problem to prove continuity of bilinear and linear forms corresponding to the classical variational formulation (Principle of Virtual Work) for linear elastostatics. (3 points)

1.3.4. Hooke's Law. For an isotropic material, the elasticities tensor depends only upon two material (Lamé) constants,

$$E_{ijkl} = \mu(\delta_{ik}\delta_{jl} + \delta_{il}\delta_{jk}) + \lambda\delta_{ij}\delta_{kl},$$

and the constitutive equations reduce to Hooke's Law:

$$\sigma_{ij} = 2\mu\epsilon_{ij} + \lambda\delta_{ij}\epsilon_{kk}.$$

Invert the constitutive law to express strains in terms of stresses,

$$\epsilon_{ij} = C_{ijkl}\sigma_{kl},$$

and derive the corresponding formula for the compliance tensor C_{ijkl}. (1 point)

1.3.5. Specialize the Lamé equations and the corresponding Principle of Virtual Work to the case of an isotropic (but not necessarily homogeneous) material. (2 points)

1.3.6. Derive the Principle of Virtual Work for the case of more general BC:

$$u_t = 0, \quad t_n = g_n \quad \text{or} \quad u_n = 0, \quad t_t = g_t,$$

where u_t, u_n denote *tangential* and *normal* components of vector u:

$$u_n = u_k n_k, \quad u_t = u - u_n n.$$

Note that u_n is a scalar, whereas u_t is a vector. Use the Fourier's Lemma argument to show formally the equivalence of classical and variational formulations. (3 points)

1.3.7. The Principle of Virtual Work involves summation in test functions v_i. Argue that the variational formulation is equivalent to a system of three variational identities where we test with just one component v_i at a time. Summing those N variational identities looks arbitrary until you consider more general BCs like those in Exercise 1.3.6. (2 points)

1.4 ▪ Variational Formulations for First Order Systems

In this section we discuss two new model problems: linear acoustics and Maxwell equations, formulated as systems of first order equations. As we will see, starting with the first order system, we open up the possibility of multiple variational formulations for the same problem. It also becomes clear which of the equations are relaxed and which are not. We also begin to use the simplified notation for the domain and boundary integrals replacing them with more compact $L^2(\Omega)$ and $L^2(\Gamma)$ symbols,

$$(u, v) = \int_\Omega u\bar{v}, \quad \langle u, v \rangle := \int_\Gamma u\bar{v}.$$

If there is a need to indicate a more specific domain of integration, we enhance the brackets with an additional symbol, e.g.,

$$(u, v)_K = \int_K u\bar{v}, \quad \langle u, v \rangle_{\Gamma_1} := \int_{\Gamma_1} u\bar{v}.$$

In the case of complex-valued problems, our default choice will be to complex-conjugate test functions, leading to the formalism of antilinear and sesquilinear forms. It goes without saying that, in case of vector- or tensor-valued functions, we use the proper dot products in place of the standard product of two numbers. In the next section, we will revisit the diffusion-convection-reaction and elasticity problems reformulated as first order systems as well.

1.4.1 ▪ Linear Acoustics Equations

The classical linear acoustics equations are obtained by linearizing the isentropic form of the compressible Euler equations expressed in terms of density ρ and velocity vector u_i, around the hydrostatic equilibrium position $\rho = \rho_0, u_i = 0$. Perturbing the solution around the equilibrium position,

$$\rho = \rho_0 + \delta\rho, \quad u_i = 0 + \delta u_i,$$

and linearizing the Euler equations (see, e.g., [56]), we obtain a system of $N + 1$ first order equations in terms of unknown perturbations of density $\delta\rho$ and velocity δu_i,

$$\begin{cases} (\delta\rho)_{,t} + \rho_0(\delta u_j)_{,j} = 0 \,, \\ \rho_0(\delta u_i)_{,t} + (\delta p)_{,i} = 0 \,, \end{cases}$$

with δp denoting the perturbation in pressure. For the isentropic[3] flow, the pressure is simply an algebraic function of density,

$$p = p(\rho) \,.$$

Linearization around the equilibrium position leads to the relation between the perturbation in density and the corresponding perturbation in pressure,

$$p = \underbrace{p(\rho_0)}_{p_0} + \frac{dp}{d\rho}(\rho_0)\delta\rho \,.$$

Here p_0 is the hydrostatic pressure, and the derivative $\frac{dp}{d\rho}(\rho_0)$ is interpreted a posteriori as the sound speed squared and denoted by c^2. Consequently, the perturbation in pressure and density are related by the simple linear equation,

$$\delta p = c^2 \delta\rho \,.$$

It is customary to express the equations of linear acoustics in terms of pressure rather than density. Dropping deltas in the notation, we obtain

$$\begin{cases} c^{-2}p_{,t} + \rho_0 u_{j,j} = 0 \,, \\ \rho_0 u_{i,t} + p_{,i} = 0 \,. \end{cases}$$

Time-harmonic equations.   Let ω denote the angular frequency. Assuming ansatz,

$$p(t, x) = e^{i\omega t}p(x), \quad u_i(t, x) = e^{i\omega t}u_i(x) \,,$$

we reduce the acoustics equations to

$$\begin{cases} c^{-2}i\omega p + \rho_0 u_{j,j} = 0 \,, \\ \rho_0 i\omega u_i + p_{,i} = 0 \,, \end{cases}$$

or, in the operator form,

$$\begin{cases} c^{-2}i\omega p + \rho_0 \operatorname{div} u = 0 \,, \\ \rho_0 i\omega u + \boldsymbol{\nabla} p = 0 \,. \end{cases}$$

---

[3]The entropy is assumed to be constant throughout the whole domain.

Nondimensionalization. Choosing reference length l_0, pressure p_0, velocity (speed) u_0, and angular frequency ω_0, we introduce nondimensional coordinates $\hat{x}_i$, pressure $\hat{p}$, velocity $\hat{u}_i$, and angular frequency $\hat{\omega}$,

$$\hat{x}_i = \frac{x_i}{l_0}, \quad \hat{p} = \frac{p}{p_0}, \quad \hat{u}_i = \frac{u_i}{u_0}, \quad \hat{\omega} = \frac{\omega}{\omega_0} \,.$$

Substituting the formulas into the equations, we get

$$\begin{cases} \dfrac{\omega_0 p_0}{c^2} i\hat{\omega}\hat{p} + \dfrac{\rho_0 u_0}{l_0} \widehat{\operatorname{div}}\hat{u} = 0 \,, \\[2mm] \rho_0 \omega_0 u_0 \, i\hat{\omega}\hat{u} + \dfrac{p_0}{l_0} \, \hat{\boldsymbol{\nabla}}\hat{p} = 0 \,. \end{cases}$$

Acoustics is a pure mechanical problem so we can choose only three independent scales (units), typically for mass (or force), length, and time (frequency in our case). For the unit of length l_0 we can choose the size of domain. For instance, if we are solving our problem in a square domain (2D), after nondimensionalization, this will be a *unit* square domain. Typically, we want the nondimensional frequency $\hat{\omega}$ to coincide with the nondimensional wave number,

$$k := \frac{\omega}{c} l_0 \,,$$

which leads to the choice of reference angular frequency, $\omega_0 = c/l_0$. Finally, we want to minimize the number of coefficients in our equations. Setting the scaling factors in the first or second equations to be equal, we obtain the relation

$$p_0 = \rho_0 c u_0 \,.$$

This means that we can choose p_0 with u_0 being derived from the equation above or, vice versa, choose u_0 and obtain p_0. Dropping the "hats," we obtain the final nondimensional equations in the form

$$\begin{cases} i\omega p + \operatorname{div} u = 0 \,, \\[2mm] i\omega u + \boldsymbol{\nabla}p = 0 \,. \end{cases} \tag{1.4.26}$$

The simplified *mathematician's acoustics equations* are thus nothing other than the properly nondimensionalized form of the original equations.

Mixed formulation I and reduction to a second order equation in terms of pressure. Eliminating the velocity, we obtain the Helmholtz equation for the pressure,

$$-\Delta p - \omega^2 p = 0 \,.$$

Having obtained the second order problem, we can proceed now with the derivation of the weak formulation, as discussed in the previous sections.

It is a little more illuminating to obtain the same variational formulation starting with the first order system. First of all, we make a clear choice in a way we treat the two equations. The equation of continuity (conservation of mass) is going to be satisfied only in the *weak sense*, i.e., we multiply it with a test function q, integrate over domain Ω, and integrate the second term by parts to obtain

$$(i\omega p, q) - (u, \boldsymbol{\nabla}q) + \langle u_n, q \rangle = 0 \quad \forall q \,,$$

where $u_n = u \cdot n = u_j n_j$ denotes the normal component of the velocity on the boundary. At this point we introduce three different BCs:

- a soft boundary Γ_p,

$$p = p_0 \,,$$

- a hard boundary Γ_u,

$$u_n = u_0 \,,$$

- and an impedance condition boundary Γ_i,

$$u_n = dp + u_0 \,,$$

where impedance constant $d > 0$.

We can now *build in* the second and third BCs into the variational formulation to obtain

$$(i\omega p, q) - (u, \nabla q) + \langle dp, q \rangle_{\Gamma_i} = -\langle u_0, q \rangle_{\Gamma_u \cup \Gamma_i} \quad \forall q : q = 0 \text{ on } \Gamma_p \,.$$

We say that *we have relaxed* the first equation. The second equation (conservation of momentum) is also multiplied with a test function v and integrated over domain Ω but *we do not integrate it by parts*,

$$(i\omega u, v) + (\nabla p, v) = 0 \quad \forall v \,.$$

If the scalar product of an L^2-function w with an arbitrary L^2 test function v vanishes,

$$(w, v) = 0 \quad \forall v \in L^2(\Omega) \,,$$

substituting $v = w$, we conclude that w must vanish a.e.,

$$\|w\|^2 = 0 \quad \Rightarrow \quad w = 0 \quad \text{a.e.}$$

Thus, except for the "a.e." symbol nothing has changed, and the equation (with $w = i\omega u + \nabla p$) is still satisfied pointwise, i.e., in *the strong sense*.

The relaxed continuity equation and strong form of the conservation of momentum equations constitute our *mixed formulation I*:

$$\begin{cases} p \in H^1(\Omega), \, p = p_0 \text{ on } \Gamma_p \,, \\ u \in L^2(\Omega) \,, \\ (i\omega p, q) - (u, \nabla q) + \langle dp, q \rangle_{\Gamma_i} = -\langle u_0, q \rangle_{\Gamma_u \cup \Gamma_i} \,, \quad q \in H^1(\Omega), \, q = 0 \text{ on } \Gamma_p \,, \\ (i\omega u, v) + (\nabla p, v) = 0 \,, \hspace{4.5cm} v \in L^2(\Omega) \,. \end{cases} \quad (1.4.27)$$

As in the previous section, choice of the energy spaces follows from the assumption on continuity (boundedness) of the sesquilinear form and the Cauchy–Schwarz inequality. Pressure p enters the formulation with gradient and, therefore, both p and ∇p must be square integrable. This leads to the assumption that $p \in H^1(\Omega)$. Similarly, no derivatives of velocity u are present in the formulation and, therefore, $u \in L^2(\Omega)$. It goes without saying that for vectors, we mean the L^2-space of vector-valued functions. Equivalently, $u \in (L^2(\Omega))^N$; see Exercise 1.4.1. It turns out that, with this choice of energy spaces, all remaining contributions to the sesquilinear form are continuous as well.

In order to fit the formulation into the abstract framework discussed in Section 1.2, we need to introduce group variables,

$$\mathsf{u} := (u, p), \quad \mathsf{v} := (v, q) \,.$$

Test and trial spaces are identical,

$$U = V := \{(v, q) \in L^2(\Omega) \times H^1(\Omega) : q = 0 \text{ on } \Gamma_p\},$$

and the antilinear and sesquilinear forms are obtained by summing up right and left sides of the formulation, respectively,

$$l(\mathsf{v}) := -\langle u_0, q \rangle_{\Gamma_u \cup \Gamma_i}$$

$$b(\mathsf{u}, \mathsf{v}) := (i\omega p, q) - (u, \boldsymbol{\nabla} q) + \langle dp, q \rangle_{\Gamma_i} + (i\omega u, v) + (\boldsymbol{\nabla} p, v).$$

The abstract formulation has the form

$$\begin{cases} \mathsf{u} \in \tilde{\mathsf{u}}_0 + U, \\ b(\mathsf{u}, \mathsf{v}) = l(\mathsf{v}), \quad \mathsf{v} \in V, \end{cases}$$

where $\tilde{\mathsf{u}}_0 = (0, \tilde{p}_0) \in L^2(\Omega) \times H^1(\Omega)$ is a finite energy lift of the BC data.

Using the (strong) conservation of momentum equation, we can represent the velocity in terms of pressure,

$$u = -\frac{1}{i\omega} \boldsymbol{\nabla} p. \tag{1.4.28}$$

In particular, the normal component of the velocity is related to the normal derivative of the pressure,

$$u_n = -\frac{1}{i\omega} \frac{\partial p}{\partial n}.$$

Multiplying $(1.4.27)_1$ with $i\omega$, and eliminating the velocity in the domain integral term using formula (1.4.28), we get the classical variational formulation of the Helmholtz equation. We can classify it as our *reduced formulation I*.

$$\begin{cases} p \in H^1(\Omega), p = p_0 \text{ on } \Gamma_p, \\ (\boldsymbol{\nabla} p, \boldsymbol{\nabla} q) - \omega^2(p, q) + i\omega \langle dp, q \rangle_{\Gamma_i} = -i\omega \langle u_0, q \rangle_{\Gamma_u \cup \Gamma_i}, \quad q \in H^1(\Omega), q = 0 \text{ on } \Gamma_p. \end{cases} \tag{1.4.29}$$

Note that we have obtained the weak formulation without introducing the second order problem at all. We have a clear understanding which of the starting equations is understood in a weak and which in a strong sense. We mention only that all these considerations can be made more precise by introducing the language of distributions and Sobolev spaces.

Mixed formulation II and reduction to a second order equation in terms of velocity.
Eliminating pressure from the first order system, we get the second order equation for the velocity,

$$-\boldsymbol{\nabla}(\operatorname{div} v) - \omega^2 u = 0.$$

As with the Helmholtz equation, we can proceed directly with the second order equation, to derive the corresponding variational formulation. But again, we prefer to work with the first order system. Keeping the conservation of mass equation in the strong form and relaxing the conservation of momentum, we obtain *mixed formulation II*.

$$\begin{cases} u_n = u_0 \text{ on } \Gamma_u, \\ i\omega(p, q) + (\operatorname{div} u, q) = 0 & \forall q, \\ i\omega(u, v) - (p, \operatorname{div} v) + \langle d^{-1} u_n, v_n \rangle_{\Gamma_i} = -\langle p_0, v_n \rangle_{\Gamma_p} + \langle d^{-1} u_0, v_n \rangle_{\Gamma_i} \\ \hspace{6cm} \forall v : v_n = 0 \text{ on } \Gamma_u. \end{cases}$$

Let us now discuss the energy setting. Unknown pressure p and test function q enter the formulation without derivatives, so $p, q \in L^2(\Omega)$. For velocity u and test function v, we employ a new energy space,

$$H(\mathrm{div}, \Omega) := \{u \in L^2(\Omega) \ : \ \mathrm{div}\, u \in L^2(\Omega)\}\,, \qquad (1.4.30)$$

where, as in the definition of $H^1(\Omega)$, divergence is understood in the distributional sense. The classical normal trace extends to a continuous operator,

$$H(\mathrm{div}, \Omega) \ni u \to u_n \in H^{-1/2}(\Gamma)$$

(see [27]), where $H^{-1/2}(\Gamma)$ is identified as the topological dual of $H^{1/2}(\Gamma)$, the trace space for $H^1(\Omega)$. This implies that terms like $\langle p_0, v_n \rangle$ can be understood in the sense of the duality pairing. Justification of term $\langle d^{-1}u_n, v_n \rangle$ is more difficult. Impedance constant d^{-1} can be factored out but the remaining term $\langle u_n, v_n \rangle$ makes sense only if we assume an additional regularity assumptions for u and/or v. The impedance BC says that, for $u_0 = 0$, normal trace u_n matches trace of p. It is thus natural to assume that the normal trace of velocity should inherit the regularity of the trace of pressure which leads to the definition of trial energy space incorporating the extra regularity assumption. If the impedance BC is applied on the *whole* boundary, $\Gamma_i = \Gamma$, we can assume

$$U := \{u \in H(\mathrm{div}, \Omega) \ : \ u_n \in H^{1/2}(\Gamma)\}\,.$$

The coupling term $\langle u_n, v_n \rangle$ can then be again understood in the sense of a duality pairing. The situation is more technical if Γ_i is a proper subset of Γ. Restriction of u_n to Γ_i, $u_n|_{\Gamma_i}$, lives still in $H^{-1/2}(\Gamma_i)$ but the corresponding dual space is no longer $H^{1/2}(\Gamma_i)$ but a more sophisticated proper subspace $\tilde{H}^{1/2}(\Gamma_i)$. This leads to the final definition of the trial energy space,

$$U := \{u \in H(\mathrm{div}, \Omega) \ : \ u_n|_{\Gamma_i} \in \tilde{H}^{1/2}(\Gamma_i)\}\,.$$

For the test space V we can keep the standard $H(\mathrm{div}, \Omega)$ space. There are two problems with this energy setting. We have lost the symmetry of the functional setting—trial and test spaces are different, which does not look natural. The second problem is more serious; the new trial energy space is more difficult to discretize in a conforming way, and trace u_n should be continuous on Γ_i. A simpler alternative is to upgrade both trial and test spaces. The Cauchy–Schwarz inequality suggests assuming the energy spaces in the form

$$U = V := \{u \in H(\mathrm{div}, \Omega) \ : \ u_n \in L^2(\Gamma_i)\}\,. \qquad (1.4.31)$$

It turns out that this space can be discretized with standard $H(\mathrm{div})$-conforming elements. Consequently, we adopt the second energy setting. The precise *mixed formulation II* now looks as follows:

$$\begin{cases} u \in V,\ u_n = u_0 \text{ on } \Gamma_u\,, \\ p \in L^2(\Omega)\,, \\ i\omega(p, q) + (\mathrm{div}\, u, q) = 0\,, & q \in L^2(\Omega)\,, \qquad (1.4.32) \\ i\omega(u, v) - (p, \mathrm{div}\, v) + \langle d^{-1}u_n, v_n \rangle_{\Gamma_i} = -\langle p_0, v_n \rangle_{\Gamma_p} + \langle d^{-1}u_0, v_n \rangle_{\Gamma_i}\,, \\ \hspace{5cm} v \in V \ : \ v_n = 0 \text{ on } \Gamma_u\,. \end{cases}$$

If we use the first equation to eliminate the pressure, we arrive at the *reduced formulation II*.

$$\begin{cases} u \in V,\ u_n = u_0 \text{ on } \Gamma_u\,, \\ (\mathrm{div}\, u, \mathrm{div}\, v) - \omega^2(u, v) + i\omega\langle d^{-1}u_n, v_n \rangle_{\Gamma_i} = -i\omega\langle p_0, v_n \rangle_{\Gamma_p} + i\omega\langle d^{-1}u_0, v_n \rangle_{\Gamma_i}\,, \\ \hspace{5cm} v \in V,\ v_n = 0 \text{ on } \Gamma_u\,. \\ \hspace{8cm} (1.4.33) \end{cases}$$

Note that we avoid using the names of Dirichlet or Neumann BCs. The condition on pressure (soft boundary) is a Dirichlet (essential) BC for reduced formulation I, but it becomes the Neumann BC in reduced formulation II. The same comment applies to the hard boundary BC.

There are two more variational formulations to go. Before we discuss them, it is convenient to introduce even more abstract notation useful for systems of first order equations. With the group variable $u := (u, p)$ in place, we introduce the operator corresponding to strong formulation (1.4.26),

$$Au := (i\omega p + \operatorname{div} u, i\omega u + \nabla p).$$

Consistently with the theory of closed operators [66], we specify the domain of the operator as

$$D(A) := \{u \in L^2(\Omega) : Au \in L^2(\Omega), p = 0 \text{ on } \Gamma_p, u = 0 \text{ on } \Gamma_u, u_n = dp \text{ on } \Gamma_i\}.$$

By assumption thus, the operator takes values in $L^2(\Omega)$. With the assumption that both p and u are L^2-functions, assumption $Au \in L^2(\Omega)$ is equivalent to conditions $\nabla p \in L^2(\Omega)$, $\operatorname{div} u \in L^2(\Omega)$. The domain of the operator can thus be written in a more concrete form:

$$D(A) := \{u = (u, p) \in H(\operatorname{div}, \Omega) \times H^1(\Omega) : p = 0 \text{ on } \Gamma_p, u = 0 \text{ on } \Gamma_u, u_n = dp \text{ on } \Gamma_i\}.$$

The adjoint operator $A^*v, v \in D(A^*)$ is defined as the operator that satisfies the equation

$$(Au, v) = (u, A^*v), \quad u \in D(A), v \in D(A^*),$$

where domain $D(A^*)$ is the maximum set for which the equality holds. Integration by parts reveals that A is formally *skew-adjoint*, $A^* = -A$, with

$$D(A^*) = \{v = (v, q) \in H(\operatorname{div}, \Omega) \times H^1(\Omega) : q = 0 \text{ on } \Gamma_p, v = 0 \text{ on } \Gamma_u, v_n = -dq \text{ on } \Gamma_i\}.$$

Note the change of sign in the impedance BC. Note also that the impedance BC implies implicitly that the velocities come actually from space V incorporating the extra regularity assumption on Γ_i.

Strong (trivial) variational formulation. Multiplying equations (1.4.26) with test functions and integrating over Ω, we obtain the *strong (trivial) variational formulation*:

$$\begin{cases} (u, p) \in H(\operatorname{div}, \Omega) \times H^1(\Omega), \\ p = p_0 \text{ on } \Gamma_p, \\ u_n = u_0 \text{ on } \Gamma_u, \\ p = du_n + u_0 \text{ on } \Gamma_i, \\ i\omega(p, q) + (\operatorname{div} u, q) = 0, \quad q \in L^2(\Omega), \\ i\omega(u, v) + (\nabla p, v) = 0, \quad v \in L^2(\Omega). \end{cases} \tag{1.4.34}$$

Using the formalism of closed operators, we can write it in a more compact form,

$$\begin{cases} u = \tilde{u}_0 + D(A), \\ (Au, v) = 0, \quad v \in L^2(\Omega), \end{cases}$$

where, as usual, $\tilde{u}_0$ is a lift of the BC data.

Ultraweak variational formulation. Integrating by parts both equations and building soft and hard BCs in, we obtain

$$\begin{cases} i\omega(p,q) - (u, \boldsymbol{\nabla} q) = -\langle u_0, q\rangle_{\Gamma_u} - \langle u_n, q\rangle_{\Gamma_i} & \forall q : q = 0 \text{ on } \Gamma_p\,, \\ i\omega(u,v) - (p, \operatorname{div} v) = -\langle p_0, v_n\rangle_{\Gamma_p} - \langle p, v_n\rangle_{\Gamma_i} & \forall v : v_n = 0 \text{ on } \Gamma_u\,. \end{cases}$$

We still have to figure out how to build in the impedance BC. This is where the adjoint operator comes in. Limiting ourselves to test functions satisfying condition $v_n = -dq$ on Γ_i, summing up the equations, and building the impedance BC in, we obtain

$$\begin{cases} \mathsf{u} \in L^2(\Omega)\,, \\ (\mathsf{u}, A^*\mathsf{v}) = -\langle u_0, q\rangle_{\Gamma_u \cup \Gamma_i} - \langle p_0, v_n\rangle_{\Gamma_p}\,, & \mathsf{v} \in D(A^*)\,. \end{cases} \tag{1.4.35}$$

Lessons learned. So, what are the lessons of this section? As we have learned, the same boundary-value problem can admit many variational formulations. One can show that all of them are simultaneously well-posed; compare [66, Section 6.6.3]. They differ in energy setting corresponding to subtle regularity assumptions on the solution. Each of them can be used as a starting point for developing a separate FE method. The functional setting will translate into convergence in different (trial) norms. The two mixed formulations along with the corresponding reduced formulations enjoy a symmetric functional setting and are eligible for the Bubnov–Galerkin method (not a must though...). The strong and UW variational formulations, with their non-symmetric functional setting, must be discretized with a Petrov–Galerkin scheme. Finally, we have introduced two more classical energy spaces: $L^2(\Omega)$ and $H(\operatorname{div}, \Omega)$.

1.4.2 ▪ Linear Elasticity Equations Revisited

We return now to the linear elasticity problem discussed in Section 1.3.2, reformulate it as a system of first order equations, and discuss possible variational formulations other than the Principle of Virtual Work.

We begin by recalling the inverse of elasticities tensor known as the *compliance tensor*,

$$\sigma_{ij} = E_{ijkl}\epsilon_{kl} \quad \Leftrightarrow \quad \epsilon_{ij} = C_{ijkl}\sigma_{kl}\,. \tag{1.4.36}$$

If the elasticities tensor represents a linear map from strains to stresses, then the compliance tensor represents its inverse. Note that both maps are defined for symmetric arguments only. The compliance tensor satisfies the same symmetry conditions as elasticities. For an isotropic material,

$$E_{ijkl} = \mu(\delta_{ik}\delta_{jl} + \delta_{il}\delta_{jk}) + \lambda\delta_{ij}\delta_{kl}\,,$$

where μ, λ denote the Lamé constants. This leads to the Hooke's Law:

$$\sigma_{ij} = 2\mu\epsilon_{ij} + \lambda\epsilon_{kk}\delta_{ij}\,.$$

The corresponding inverse formula takes the form

$$\epsilon_{ij} = \frac{1}{2\mu}\sigma_{ij} - \frac{\lambda}{2\mu(2\mu + N\lambda)}\sigma_{kk}\delta_{ij}$$

or

$$\epsilon_{ij} = \frac{1}{2\mu}\sigma_{ij} - \frac{1}{2\mu(\frac{2\mu}{\lambda} + N)}\sigma_{kk}\delta_{ij}\,.$$

The two laws behave differently when attempting to pass to the incompressible limit, $\lambda \to \infty$. Whereas the norm of elastictities tensor blows up to infinity, the compliance law converges seamlessly to

$$\epsilon_{ij} = \frac{1}{2\mu} \underbrace{\left(\sigma_{ij} - \frac{1}{N} \sigma_{kk} \delta_{ij} \right)}_{=:\sigma_{ij}^{\text{dev}}} .$$

The norm of the compliance tensor stays bounded and, in the limit, the strain depends entirely upon the stress deviator σ_{ij}^{dev} only.

The antisymmetric part of the displacement gradient is identified as the *linearized rigid body motion*:

$$r_{ij} := \frac{1}{2}(u_{i,j} - u_{j,i}) .$$

Combining the compliance law with the definition of r_{ij}, we get

$$C_{ijkl}\sigma_{kl} = u_{i,j} - r_{ij} .$$

Note that the equation above contains the definition of tensor r_{ij}. Indeed, it suffices to take the nonsymmetric part of both sides of the equation. Note also that, even if we extend the validity of the equation to arbitrary (nonnecessary symmetric) tensors σ_{kl}, symmetry condition $C_{ijkl} = C_{ijlk}$ implies that the left-hand side "sees" only the symmetric part of the stress.

The time-harmonic version of the elastodynamics problem can now be formulated as a system of first order equations:

$$\begin{cases} -\sigma_{ij,j} - \rho\omega^2 u_i = f_i & \text{in } \Omega , \\ C_{ijkl}\sigma_{kl} - u_{i,j} + r_{ij} = 0 & \text{in } \Omega , \\ \sigma_{ij} - \sigma_{ji} = 0 & \text{in } \Omega , \\ u_i = 0 & \text{on } \Gamma_u , \\ \sigma_{ij}n_j = 0 & \text{on } \Gamma_t . \end{cases}$$

All unknowns—displacements u_i, stresses σ_{ij}, and infinitesimal rotation tensor r_{ij}—are complex-valued, ω denotes the angular frequency, and ρ is the density of mass. The first system represents conservation of linear momentum, the second represents a constitutive equation with definition of r_{ij} combined, and the third one (symmetry of stress) derives from the conservation of angular momentum. In order to simplify the discussion, we stick with homogeneous BCs only. Nonhomogeneous BCs can always be taken into account by means of finite energy lifts.

We switch now to the absolute notation.

$$\begin{cases} -\operatorname{div} \sigma - \rho\omega^2 u = f & \text{in } \Omega , \\ C\sigma - \boldsymbol{\nabla}u + r = 0 & \text{in } \Omega , \\ \sigma - \sigma^T = 0 & \text{in } \Omega , \\ u = 0 & \text{on } \Gamma_u , \\ \sigma n = 0 & \text{on } \Gamma_t . \end{cases} \qquad (1.4.37)$$

In the following discussion we will restrict ourselves to $N = 3$.

Strong (trivial) variational formulations. Multiplying the equations with test functions v, τ_{ij} and antisymmetric tensors $s = -s^T$, and integrating over Ω, we obtain

$$\begin{cases} -(\operatorname{div}\sigma, v) - \omega^2(\rho u, v) = (f, v)\,, \\ (C\sigma, \tau) - (\boldsymbol{\nabla}u, \tau) + (r, \tau) = 0\,, \\ \qquad\qquad\qquad (\sigma, s) = 0\,, \qquad s = -s^T\,, \end{cases} \qquad (1.4.38)$$

with group unknown $\mathsf{u} := (u, \sigma, r)$,

$$u \in H^1(\Omega)^3 : u = 0 \quad \text{on } \Gamma_u\,,$$
$$\sigma \in H(\operatorname{div}, \Omega)^3 : \sigma n = 0 \quad \text{on } \Gamma_t\,,$$
$$r \in L^2(\Omega)^3\,,$$

and group test function $\mathsf{v} := (v, \tau, s)$,

$$v \in L^2(\Omega)^3\,,$$
$$\tau \in L^2(\Omega)^{3\times3}\,,$$
$$s = -s^T \in L^2(\Omega)^3\,.$$

By delegating the symmetry of stress to a separate equation, we are able to look for the stresses in (a larger) space $H(\operatorname{div}, \Omega)^3$ consisting of just three copies of the standard $H(\operatorname{div}, \Omega)$ space. An alternate, strong imposition of the symmetry leads to a smaller energy space,

$$H^{\mathrm{sym}}(\operatorname{div}, \Omega) := \{\sigma_i. \in H(\operatorname{div}, \Omega), i = 1, \ldots, 3 : \sigma_{ij} = \sigma_{ji}\}\,,$$

which is much more difficult to discretize.

If we are not interested in r_{ij}, we can eliminate it by testing in the second equation with symmetric tensors $\tau = \tau^T$ only:

$$\begin{cases} -(\operatorname{div}\sigma, v) - \omega^2(\rho u, v) = (f, v)\,, \\ \quad (C\sigma, \tau) - (\boldsymbol{\nabla}u, \tau) = 0\,, \qquad \tau = \tau^T\,, \\ \qquad\qquad\quad (\sigma, s) = 0\,, \qquad s = -s^T\,, \end{cases} \qquad (1.4.39)$$

with group unknown $\mathsf{u} := (u, \sigma)$,

$$u \in H^1(\Omega)^3 : u = 0 \quad \text{on } \Gamma_u\,,$$
$$\sigma \in H(\operatorname{div}, \Omega)^3 : \sigma n = 0 \quad \text{on } \Gamma_t\,,$$

and group test function $\mathsf{v} := (v, \tau, s)$,

$$v \in L^2(\Omega)^3\,,$$
$$\tau = \tau^T \in L^2(\Omega)^6\,,$$
$$s = -s^T \in L^2(\Omega)^3\,.$$

Both formulations (1.4.38) and (1.4.39) use a nonsymmetric functional setting and cannot be approximated with the standard Bubnov–Galerkin method.

Mixed variational formulation I. Relaxing[4] the momentum equations, we obtain

$$
\begin{cases}
(\sigma, \nabla v) - \omega^2(\rho u, v) = (f, v) & \text{relaxed}, \\
(C\sigma, \tau) - (\nabla u, \tau) + (r, \tau) = 0, \\
\qquad\qquad (\sigma, s) = 0, & s = -s^T,
\end{cases}
\tag{1.4.40}
$$

with group unknown $\mathsf{u} := (u, \sigma, r)$,

$$
\begin{aligned}
u &\in H^1(\Omega)^3 &&: u = 0 \text{ on } \Gamma_u, \\
\sigma &\in L^2(\Omega)^{3\times 3}, \\
r &\in L^2(\Omega)^3,
\end{aligned}
$$

and group test function $\mathsf{v} := (v, \tau, s)$,

$$
\begin{aligned}
v &\in H^1(\Omega)^3 &&: v = 0 \text{ on } \Gamma_u, \\
\tau &\in L^2(\Omega)^{3\times 3}, \\
s &\in L^2(\Omega)^3.
\end{aligned}
$$

This time, the functional setting is symmetric.

As before, we can test only with symmetric τ and eliminate r,

$$
\begin{cases}
(\sigma, \nabla v) - \omega^2(\rho u, v) = (f, v) & \text{relaxed}, \\
(C\sigma, \tau) - (\nabla u, \tau) = 0
\end{cases}
\tag{1.4.41}
$$

with group unknown $\mathsf{u} := (u, \sigma)$,

$$
\begin{aligned}
u &\in H^1(\Omega)^3 : u = 0 \text{ on } \Gamma_u, \\
\sigma &= \sigma^T \in L^2(\Omega)^6,
\end{aligned}
$$

and group test function $\mathsf{v} := (v, \tau)$,

$$
\begin{aligned}
v &\in H^1(\Omega)^3 : v = 0 \text{ on } \Gamma_u, \\
\tau &= \tau^T \in L^2(\Omega)^6.
\end{aligned}
$$

As before, we have a symmetric functional setting. Finally, we can reverse to the original form of the constitutive law and eliminate the stress to formulate the problem entirely in terms of the displacement vector.

Reduced variational formulation I. We have arrived at the classical Principle of Virtual Work,

$$
(E\nabla u, \nabla v) - \omega^2(\rho u, v) = (f, v),
\tag{1.4.42}
$$

with unknown

$$
u \in H^1(\Omega)^3 : u = 0 \text{ on } \Gamma_u
$$

and test function

$$
v \in H^1(\Omega)^3 : v = 0 \text{ on } \Gamma_u.
$$

[4]Integrating by parts and building the corresponding BC in.

A reminder. $f = 0$ in Ω implies $(f, v) = 0$ for every $v \in L^2(\Omega)$. Conversely, if the condition is satisfied, selecting $v = f$, we conclude that $\|f\| = 0 \Rightarrow f = 0$ a.e. As we revert from the strong (nonrelaxed) form of an equation to its pointwise version, we understand it always in the L^2-sense, i.e., the equation is satisfied only *a.e.* in Ω.

Mixed variational formulation II. We get another symmetric functional setting by relaxing the constitutive equations,

$$
\begin{cases}
-(\operatorname{div}\sigma, v) - \omega^2(\rho u, v) = (f, v), \\
(C\sigma, \tau) + (u, \operatorname{div}\tau) + (r, \tau) = 0 \qquad \text{relaxed}, \\
\qquad\qquad\qquad (\sigma, s) = 0, \qquad s = -s^T,
\end{cases}
\tag{1.4.43}
$$

with group unknown $\mathsf{u} := (u, \sigma, r)$,

$$
u \in L^2(\Omega)^3,
$$
$$
\sigma \in H(\operatorname{div}, \Omega)^3 : \sigma n = 0 \quad \text{on } \Gamma_t,
$$
$$
r \in L^2(\Omega)^3,
$$

and group test function $\mathsf{v} := (v, \tau, s)$,

$$
v \in L^2(\Omega)^3,
$$
$$
\tau \in H(\operatorname{div}, \Omega)^3 : \tau n = 0 \quad \text{on } \Gamma_t,
$$
$$
s \in L^2(\Omega)^3.
$$

Reduced variational formulation II. For $\omega \neq 0$, we can use the first equation to eliminate u to obtain another variational formulation with a symmetric functional setting,

$$
\begin{cases}
(C\sigma, \tau) - \omega^{-2}(\rho^{-1}\operatorname{div}\sigma, \operatorname{div}\tau) + (r, \tau) = \omega^{-2}(\rho^{-1}f, \operatorname{div}\tau), \\
\qquad\qquad\qquad (\sigma, s) = 0, \qquad\qquad s = -s^T,
\end{cases}
\tag{1.4.44}
$$

with group unknown $\mathsf{u} := (\sigma, r)$,

$$
\sigma \in H(\operatorname{div}, \Omega)^3 : \sigma n = 0 \quad \text{on } \Gamma_t,
$$
$$
r \in L^2(\Omega)^3,
$$

and group test function $\mathsf{v} := (\tau, s)$,

$$
\tau \in H(\operatorname{div}, \Omega)^3 : \tau n = 0 \quad \text{on } \Gamma_t,
$$
$$
s \in L^2(\Omega)^3.
$$

Ultraweak variational formulation. Our final formulation is based on relaxing both equations,

$$
\begin{cases}
(\sigma, \boldsymbol{\nabla}v) - \omega^2(\rho u, v) = (f, v) \qquad \text{relaxed}, \\
(C\sigma, \tau) + (u, \operatorname{div}\tau) + (r, \tau) = 0 \qquad \text{relaxed}, \\
\qquad\qquad\qquad (\sigma, s) = 0, \qquad s = -s^T,
\end{cases}
\tag{1.4.45}
$$

with group unknown u := (u, σ, r),

$$u \in L^2(\Omega)^3 \,,$$
$$\sigma \in L^2(\Omega)^{3\times 3} \,,$$
$$r \in L^2(\Omega)^3 \,,$$

and group test function v := (v, τ, s),

$$v \in H^1(\Omega)^3 \quad : \quad v = 0 \quad \text{on } \Gamma_u \,,$$
$$\tau \in H(\text{div}, \Omega)^3 : \tau n = 0 \quad \text{on } \Gamma_t \,,$$
$$s \in L^2(\Omega)^3 \,.$$

Clearly, we have an unsymmetric functional setting. Enforcing the symmetry of the L^2 stress tensor is now easy, and we may eliminate the last equation to obtain a reduced form of the UW formulation.

$$\begin{cases} (\sigma, \boldsymbol{\nabla} v) - \omega^2(\rho u, v) = (f, v) & \text{relaxed}, \\ (C\sigma, \tau) + (u, \text{div } \tau) + (r, \tau) = 0 & \text{relaxed} \end{cases} \tag{1.4.46}$$

with group unknown u := (u, σ, r),

$$u \in L^2(\Omega)^3 \,,$$
$$\sigma = \sigma^T \in L^2(\Omega)^6 \,,$$
$$r \in L^2(\Omega)^3 \,,$$

and group test function v := (v, τ),

$$v \in H^1(\Omega)^3 \quad : v = 0 \quad \text{on } \Gamma_u \,,$$
$$\tau \in H(\text{div}, \Omega)^3 : \tau n = 0 \quad \text{on } \Gamma_t \,.$$

As we can see, we have a multitude of possible variational formulations, all involving the H^1, $H(\text{div})$, and L^2 energy spaces. They can accommodate more or less regular solutions corresponding to loads of specific regularity. One can show that the sesquilinear forms corresponding to the different formulations simultaneously do or do not satisfy the inf-sup conditions; see [53]. Each formulation may give rise to a separate FE method for elasticity with the numerical solution converging in the norm corresponding to the specific functional setting.

Incompressible limit. The reduced variational formulation (1.4.42) is based on the original version of the constitutive law, and it loses its stability in the incompressible limit when $\lambda \to \infty$. All remaining formulations are based on the compliance form and stay valid for $\lambda = \infty$. This suggests that the FE methods based on these formulations will have a chance to avoid the so-called *volumetric locking*.

The Stokes Problem

The best known formulation that remains valid in the incompressible limit is formulated in terms of displacement and just one additional scalar-valued unknown—the *pressure*. We start by recalling the Principle of Virtual Work for the isotropic elasticity (compare Exercise 1.3.5),

$$\begin{cases} u \in H^1(\Omega)^N, \ u = 0 \text{ on } \Gamma_1 \,, \\ \int_\Omega [\mu(\boldsymbol{\nabla} u + \boldsymbol{\nabla}^T u) \boldsymbol{\nabla} v + \lambda \, \text{div } u \, \text{div } v] = \int_\Omega fv + \int_{\Gamma_2} gv \quad v \in H^1(\Omega)^N \ : v = 0 \text{ on } \Gamma_1 \,. \end{cases}$$

Introducing a new variable, pressure $p = \lambda \operatorname{div} u$, into the equation and imposing the definition in the weak form, we obtain the formulation

$$
\begin{cases}
u \in H^1(\Omega)^N, \ u = 0 \text{ on } \Gamma_1, \ p \in L^2(\Omega), \\[2mm]
\displaystyle \int_\Omega \mu(\boldsymbol{\nabla} u + \boldsymbol{\nabla}^T u)\boldsymbol{\nabla} v + \int_\Omega p \operatorname{div} v = \int_\Omega fv + \int_{\Gamma_2} gv \\[2mm]
\hspace{4cm} v \in H^1(\Omega)^N \ : \ v = 0 \text{ on } \Gamma_1, \\[2mm]
\displaystyle \int_\Omega \operatorname{div} u\, q, \hspace{1cm} -\frac{1}{\lambda}\int_\Omega p q = 0, \hspace{1cm} q \in L^2(\Omega).
\end{cases}
\tag{1.4.47}
$$

Note that we have imposed the definition of pressure in the compliance form (divided by λ). Passing with $\lambda \to \infty$, we obtain the variational formulation for the *Stokes problem*,

$$
\begin{cases}
u \in H^1(\Omega)^N, \ u = 0 \text{ on } \Gamma_1, \ p \in L^2(\Omega), \\[2mm]
\displaystyle \int_\Omega \mu(\boldsymbol{\nabla} u + \boldsymbol{\nabla}^T u)\boldsymbol{\nabla} v + \int_\Omega p \operatorname{div} v = \int_\Omega fv + \int_{\Gamma_2} gv, \\[2mm]
\hspace{4cm} v \in H^1(\Omega)^N \ : \ v = 0 \text{ on } \Gamma_1, \\[2mm]
\displaystyle \int_\Omega \operatorname{div} u\, q = 0, \hspace{1cm} q \in L^2(\Omega).
\end{cases}
\tag{1.4.48}
$$

In the case of pure kinematic BCs, i.e., $\Gamma_2 = \emptyset$,

$$
\int_\Omega \mu u_{j,i} v_{i,j} = -\int_\Omega \mu u_{j,ij} v_i = -\int_\Omega \mu \underbrace{(\operatorname{div} u)_{,i}}_{=0} v_i = 0.
$$

Pressure p is then determined up to a constant only. To ensure uniqueness, we have to seek pressure in the quotient space $L^2(\Omega)/\mathbb{R}$ or, equivalently, impose an additional scaling condition, e.g., $\int_\Omega p = 0$. The final formulation looks as follows:

$$
\begin{cases}
u \in H_0^1(\Omega)^N, \ p \in L_0^2(\Omega), \\[2mm]
\displaystyle \int_\Omega \mu \boldsymbol{\nabla} u\, \boldsymbol{\nabla} v + \int_\Omega p \operatorname{div} v = \int_\Omega fv, \hspace{1cm} v \in H_0^1(\Omega)^N, \\[2mm]
\displaystyle \int_\Omega \operatorname{div} u\, q = 0, \hspace{1cm} q \in L_0^2(\Omega),
\end{cases}
\tag{1.4.49}
$$

where

$$
H_0^1(\Omega) := \{ u \in H^1(\Omega) \ : \ u = 0 \text{ on } \Gamma \},
$$

$$
L_0^2(\Omega) := \left\{ q \in L^2(\Omega) \ : \ \int_\Omega q = 0 \right\}.
$$

Remark 1.5. One can define the pressure in terms of the axiatoric (volumetric) part of the stress, $p = -\sigma_{ii}/N$. This leads to a slightly more complicated formulation than (1.4.47). In the incompressible limit though, both formulations reduce to (1.4.48).

1.4.3 ▪ Maxwell Equations

For a short introduction to Maxwell equations, we refer the reader to [28].

We shall consider the time-harmonic Maxwell equations:

- Faraday's Law,

$$\frac{1}{\mu}\boldsymbol{\nabla}\times E = -\frac{1}{\mu}K^{\text{imp}} - i\omega H\,, \tag{1.4.50}$$

- and Ampère's Law,

$$\boldsymbol{\nabla}\times H = J^{\text{imp}} + \sigma E + i\omega\epsilon E\,. \tag{1.4.51}$$

Here ϵ, μ, σ denote the material constants—permittivity, permeability and conductivity—and J^{imp} and K^{imp} stand for a prescribed impressed electric or magnetic current, respectively. The system above can be assumed be already in a nondimensional form; see Exercise 1.4.3. We shall assume that all material constants are real and bounded, with permittivity and permeability bounded away from zero,

$$\epsilon(x) \geq \epsilon_0 > 0\,, \quad \mu(x) \geq \mu_0 > 0\,. \tag{1.4.52}$$

As for the acoustics equations, we can develop six different variational formulations; see Exercise 1.4.4. We summarize here the two classical (reduced) formulations, in terms of either electric or magnetic field alone. Depending upon the choice, one of the equations is going to be satisfied in a weak sense, and the other one in the strong sense. If we choose to solve for the electric field, we multiply Ampère's Law with $-i\omega$, then with a test function F, integrate over Ω and integrate by parts to obtain

$$(-i\omega H, \boldsymbol{\nabla}\times F) - ((\omega^2\epsilon - i\omega\sigma)E, F) - i\omega\langle n\times H, F\rangle = -i\omega(J^{\text{imp}}, F) \quad \forall F\,. \tag{1.4.53}$$

We introduce now the BCs:

- perfectly conducting boundary on Γ_E,

$$n\times E = n\times E_0\,,$$

- prescribed electric surface current on Γ_H,

$$n\times H = J_S^{\text{imp}} := n\times H_0\,,$$

- an impedance BC on Γ_i,

$$n\times H + dE_t = J_S^{\text{imp}}\,. \tag{1.4.54}$$

Here $E_t = -n\times(n\times E)$ stands for the tangential component of E, d is a prescribed impedance, and J_S^{imp} is a prescribed electric surface current. Notice that the impressed surface current is tangent to the boundary.

Introducing the BCs into (1.4.53), we obtain

$$(-i\omega H, \boldsymbol{\nabla}\times F) - ((\omega^2\epsilon - i\omega\sigma)E, F) + i\omega\langle dE_t, F\rangle_{\Gamma_i} = -i\omega(J^{\text{imp}}, F) + i\omega\langle J_S^{\text{imp}}, F\rangle_{\Gamma_H\cup\Gamma_i}$$

$$\forall F : n\times F = 0 \text{ on } \Gamma_E\,.$$

Notice that $E_t F = E_t F_t$ and $J_S^{\text{imp}} F = J_S^{\text{imp}} F_t$.

The final variational formulation is obtained by using the Faraday equation to eliminate the magnetic field. We obtain

$$\begin{cases} n\times E = n\times E_0 \text{ on } \Gamma_E\,, \\[2mm] \left(\dfrac{1}{\mu}\boldsymbol{\nabla}\times E, \boldsymbol{\nabla}\times F\right) - ((\omega^2\epsilon - i\omega\sigma)E, F) + i\omega\langle dE_t, F\rangle_{\Gamma_i} \\[2mm] = -i\omega(J^{\text{imp}}, F) - \left(\dfrac{1}{\mu}K^{\text{imp}}, \boldsymbol{\nabla}\times F\right) + i\omega\langle J_S^{\text{imp}}, F\rangle_{\Gamma_H\cup\Gamma_i} \\[2mm] \hspace{3cm} \forall F : n\times F = 0 \text{ on } \Gamma_E\,. \end{cases}$$

Well-posedness considerations lead to a new energy space

$$H(\operatorname{curl}, \Omega) := \{E \in L^2(\Omega) \,:\, \boldsymbol{\nabla} \times E \in L^2(\Omega)\}\,. \tag{1.4.55}$$

The new space comes with two trace operators,

$$\gamma_t : \; H(\operatorname{curl}, \Omega) \ni E \to E_t \in H^{-1/2}(\operatorname{curl}, \Gamma)\,,$$
$$\gamma_t^{\perp} : H(\operatorname{curl}, \Omega) \ni E \to n \times E_t \in H^{-1/2}(\operatorname{div}, \Gamma) = (H^{-1/2}(\operatorname{curl}, \Gamma))'\,. \tag{1.4.56}$$

The precise definition of trace operators and trace spaces is quite involved [27]. Finally, a discussion on the impedance BCs fully analogous to the acoustics problem leads to the extra regularity assumption built into the definition of the proper energy space,

$$Q := \{E \in H(\operatorname{curl}, \Omega) \,:\, E_t \in \tilde{H}^{-1/2}(\operatorname{div}, \Gamma_i)\}\,,$$

with a properly defined space $\tilde{H}^{-1/2}(\operatorname{div}, \Gamma_i)$. Similarly to the acoustics problem, an easier alternative uses L^2-space,

$$Q := \{E \in H(\operatorname{curl}, \Omega) \,:\, E_t \in L^2(\Gamma_i)\}\,. \tag{1.4.57}$$

The assumption makes the term $\langle E_t, F_t \rangle_{\Gamma_i}$ legitimate. The final precise formulation looks as follows:

$$\begin{cases} E \in Q,\, n \times E = n \times E_0 \text{ on } \Gamma_E\,, \\ \left(\dfrac{1}{\mu}\boldsymbol{\nabla} \times E, \boldsymbol{\nabla} \times F\right) - ((\omega^2\epsilon - i\omega\sigma)E, F) + i\omega\langle dE_t, F\rangle_{\Gamma_i} \\ = -i\omega(J^{\mathrm{imp}}, F) - \left(\dfrac{1}{\mu}K^{\mathrm{imp}}, \boldsymbol{\nabla} \times F\right) + i\omega\langle J_S^{\mathrm{imp}}, F\rangle_{\Gamma_H \cup \Gamma_i}\,, \\ \hspace{5cm} F \in Q,\, n \times F = 0 \text{ on } \Gamma_E\,. \end{cases} \tag{1.4.58}$$

Formulation in terms of the magnetic field. If we choose to work with the magnetic field, we treat the Faraday equation in the weak form. Since permeability μ may be a function of x, we first multiply the equation with μ, and only then test it with a test function F to obtain

$$(E, \boldsymbol{\nabla} \times F) + i\omega(\mu H, F) + \langle n \times E, F\rangle = -(K^{\mathrm{imp}}, F) \quad \forall F\,. \tag{1.4.59}$$

We discuss now the BCs:

- prescribed electric surface current on Γ_H,

$$n \times H = n \times H_0\,;$$

- perfectly conducting boundary on Γ_E, i.e., a prescribed magnetic surface current:

$$n \times E = -K_S^{\mathrm{imp}} := n \times E_0\,,$$

- impedance BC on Γ_i:

$$n \times E - \frac{1}{d}H_t = \frac{1}{d}n \times J_S^{\mathrm{imp}} =: -K_S^{\mathrm{imp}}\,.$$

Notice that the definition of the Dirichlet or Neumann part of the boundary depends upon the formulation. The Dirichlet data for the E-formulation has now become Neumann data, and vice versa. The new form of the impedance BC has been obtained by multiplying (1.4.54) on the left by $n\times$ and dividing by impedance constant d. Substituting the BC data into the boundary term in formulation (1.4.59), and restricting ourselves to test functions satisfying the homogeneous Dirichlet BC we get

$$(E, \boldsymbol{\nabla} \times F) + i\omega(\mu H, F) + \left\langle \frac{1}{d} H_t, F \right\rangle_{\Gamma_i} = -(K^{\mathrm{imp}}, F) + \langle K_S^{\mathrm{imp}}, F \rangle_{\Gamma_E \cup \Gamma_i}$$
$$\forall F \; : \; n \times F = 0 \text{ on } \Gamma_H \,.$$

The final variational formulation is obtained by using the Ampère's Law to eliminate the electric field:

$$\begin{cases} H \in Q, \; n \times H = n \times H_0 \text{ on } \Gamma_H \,, \\[2mm] \left(\dfrac{1}{i\omega\epsilon + \sigma} \boldsymbol{\nabla} \times H, \boldsymbol{\nabla} \times F \right) + i\omega(\mu H, F) + \left\langle \dfrac{1}{d} H_t, F \right\rangle_{\Gamma_i} \\[3mm] = -(K^{\mathrm{imp}}, F) + \left(\dfrac{1}{i\omega\epsilon + \sigma} J^{\mathrm{imp}}, \boldsymbol{\nabla} \times F \right) + \langle K_S^{\mathrm{imp}}, F \rangle_{\Gamma_E \cup \Gamma_i} \,, \\[3mm] \hspace{4cm} F \in Q, \; n \times F = 0 \text{ on } \Gamma_H \,. \end{cases} \quad (1.4.60)$$

The energy space Q incorporates again the extra regularity condition on the impedance boundary,

$$Q := \{ H \in H(\mathrm{curl}, \Omega) \; : \; H_t \in L^2(\Gamma_I) \} \,.$$

1.4.4 ▪ Maxwell Equations: A Deeper Look

The story behind Maxwell's equations goes much deeper behind the need for a new energy space $H(\mathrm{curl}, \Omega)$. Complete (time harmonic) Maxwell's equations include not only the Faraday and Ampère Laws but also the two Gauss laws and the conservation of (free) charge equation.

$$\begin{cases} \boldsymbol{\nabla} \times E = -i\omega(\mu H) & \text{(Faraday's Law)}\,, \\[2mm] \boldsymbol{\nabla} \times H = J^{\mathrm{imp}} + \underbrace{\sigma E}_{J} + i\omega(\epsilon E) & \text{(Ampère's Law)}\,, \\[2mm] \boldsymbol{\nabla} \cdot (\mu H) = 0 & \text{(Gauss's Magnetic Law)}\,, \\[2mm] \boldsymbol{\nabla} \cdot (\epsilon E) = \rho^{imp} + \rho & \text{(Gauss's Electric Law)}\,, \\[2mm] i\omega\rho + \boldsymbol{\nabla} \cdot \boldsymbol{J} = 0 & \text{(conservation of charge)}\,. \end{cases} \quad (1.4.61)$$

To simplify the presentation, we have assumed $K^{\mathrm{imp}} = 0$. We have a total of seven scalar unknowns: three components of E, H each and ρ, and a total of nine scalar equations. Obviously, the equations are linearly dependent. To simplify the discussion, we can eliminate the free charge density by combining the last two equations into one (we will call it the "continuity equation"),

$$\begin{cases} \boldsymbol{\nabla} \times E = -i\omega(\mu H) & \text{(Faraday's Law)}\,, \\[2mm] \boldsymbol{\nabla} \times H = J^{\mathrm{imp}} + \underbrace{\sigma E}_{J} + i\omega(\epsilon E) & \text{(Ampère's Law)}\,, \\[2mm] \boldsymbol{\nabla} \cdot (\mu H) = 0 & \text{(Gauss's Magnetic Law)}\,, \\[2mm] -i\omega\rho^{imp} + \boldsymbol{\nabla} \cdot \boldsymbol{J} + i\omega\boldsymbol{\nabla} \cdot (\epsilon E) = 0 & \text{(continuity equation)}\,. \end{cases}$$
$$(1.4.62)$$

The algebraic dependence structure is now clearly visible. Gauss's Magnetic Law is obtained by applying the divergence operator to both sides of Faraday's Law, and the continuity equation is obtained by taking the divergence of Ampère's Law. The last two equations are thus automatically satisfied once the first two hold. Note that once the electric field E is known, either Gauss's Electric Law or the conservation of charge equation can be used to compute the free charge density ρ. Notice also that the prescribed impressed current and charge must be compatible with each other (satisfy the conservation of charge equation).

Critical to the discretization of Maxwell equations is the fact that this automatic satisfaction of Gauss's Magnetic Law and the continuity equations carries over to the weak form of the equations, and then to the discrete level as well.

We shall focus on the formulation (1.4.58) in terms of electric field E. Analogous results hold for the other formulation as well. First of all, once the electric field is known, the corresponding magnetic field is computed using the strong form of Faraday's Law:

$$-i\mu\omega H = \boldsymbol{\nabla} \times E\,.$$

Taking the divergence of both sides, we verify easily the Gauss's Magnetic Law.

In order to recover the continuity equation from variational formulation (1.4.58), we employ a special test function $F = \boldsymbol{\nabla}q$ where $q \in H^1(\Omega)$, $q = 0$ on Γ_E to obtain

$$-((\omega^2\epsilon - i\omega\sigma)E, \boldsymbol{\nabla}q) + i\omega\langle dE_t, \boldsymbol{\nabla}q\rangle_{\Gamma_i} = -i\omega(J^{\mathrm{imp}}, \boldsymbol{\nabla}q) + i\omega(J_S^{imp}, \boldsymbol{\nabla}q)_{\Gamma_H \cup \Gamma_i} \qquad \forall q\,. \tag{1.4.63}$$

The equation represents not only a weak form of the continuity equation but also additional (automatically satisfied) BCs on Γ_H and Γ_i.

The critical point here is the fact that we *could make the substitution* $F = \boldsymbol{\nabla}q$, i.e., that the gradients $\boldsymbol{\nabla}q$ live in the energy space $H(\mathrm{curl}, \Omega)$.

1.4.5 ▪ Stabilized Formulation

Related to the implicit satisfaction of the continuity equation is the concept of the so-called stabilized formulation [38]. For simplicity of presentation, we will restrict ourselves to the case of $\Gamma_i = \emptyset$, $E_0 = 0$, and $\sigma = 0$. We impose the implicitly satisfied equation (1.4.63) as an additional constraint and introduce the corresponding Lagrange multiplier p. The new formulation looks as follows:

$$\begin{cases} E \in H(\mathrm{curl}, \Omega),\ n \times E = 0 \text{ on } \Gamma_E,\ p \in H^1(\Omega),\ p = 0 \text{ on } \Gamma_E\,, \\[2mm] \left(\dfrac{1}{\mu}\boldsymbol{\nabla} \times E, \boldsymbol{\nabla} \times F\right) - \omega^2(\epsilon E, F) - \omega^2(\epsilon\boldsymbol{\nabla}p, F) = -i\omega(J^{\mathrm{imp}}, F) + i\omega\langle J_S^{\mathrm{imp}}, F\rangle_{\Gamma_H}\,, \\[2mm] \hspace{5cm} F \in H(\mathrm{curl}, \Omega),\ n \times F = 0 \text{ on } \Gamma_E \\[2mm] -\omega^2(\epsilon E, \boldsymbol{\nabla}q) = -i\omega(J^{\mathrm{imp}}, \boldsymbol{\nabla}q) + i\omega\langle J_S^{\mathrm{imp}}, \boldsymbol{\nabla}q\rangle_{\Gamma_H}\,, \\[2mm] \hspace{6cm} q \in H^1(\Omega),\ q = 0 \text{ on } \Gamma_E\,. \end{cases} \tag{1.4.64}$$

The name *stabilized* comes from the fact that for $\sigma = 0$, we can divide the second equation by ω and drop the ω factor in the Lagrange multiplier term as well. The stabilized formulation then exhibits better stability properties than the original formulation with $\omega \to 0$. In the case when the right-hand side in the second equation vanishes (an additional assumption on the data), we can drop the whole factor ω^2 in both the second equation and the Lagrange multiplier term. Contrary to the original formulation, the stabilized formulation remains then uniformly stable as $\omega \to 0$. See [38] for details.

The added constraint was implicitly satisfied by the solution to the original problem, which suggests that the Lagrange multiplier (representing a reaction to the imposed constraint) should vanish. Indeed, testing the first equation with $F = \nabla p$, and utilizing the second equation, we obtain

$$\omega^2 (\epsilon \nabla p, \nabla p) = 0 \quad \Rightarrow \quad \nabla p = 0$$

which, in the presence of the BC on Γ_E, implies $p = 0$. The two variational problems are thus equivalent. Note that the two variational formulations are equivalent also on the discrete level, provided the discrete space for Lagrange multiplier p is such that the gradient maps it into a subspace of the discrete $H(\text{curl})$-conforming space. We arrive at the need of discrete spaces forming the exact sequence to be discussed in the next chapter.

The stabilized formulation has the structure of a mixed problem and analyzing its well-posedness and convergence of Galerkin discretization is somehow easier than for the original formulation.

Exercises

1.4.1. Explain why the space of vector-valued L^2-functions,

$$\boldsymbol{L}^2(\Omega) := \left\{ \boldsymbol{u} : \Omega \to \mathbb{C}^N \ : \ \int_\Omega |\boldsymbol{u}|^2 < \infty \right\},$$

is isomorphic and isometric with N copies of scalar-valued functions,

$$(L^2(\Omega))^N.$$

(1 point)

1.4.2. Write down explicitly trial and test spaces, and formulas for sesquilinear and antilinear forms for all six variational formulations for the acoustics problem. Assume homogeneous essential BCs to avoid affine spaces. (1 point)

1.4.3. Discuss nondimensionalization of time-harmonic Maxwell equations. How many independent units are involved? (2 points)

1.4.4. Consider the Faraday and Ampère Laws,

$$\nabla \times E = -i\omega\mu H \qquad \qquad \text{(Faraday's Law)},$$

$$\nabla \times H = J^{\text{imp}} + \sigma E + i\omega\epsilon E \qquad \text{(Ampère's Law)},$$

accompanied with BCs:

$$n \times E = n \times E_0 \qquad \text{on } \Gamma_E,$$

$$n \times H = n \times H_0 \qquad \text{on } \Gamma_H,$$

$$n \times H + dE_t = n \times H_0 \qquad \text{on } \Gamma_i.$$

Proceed along exactly the same lines as for acoustics equations to derive mixed, reduced, trivial, and UW variational formulations for Maxwell equations. (5 points)

1.4.5. Integration by parts formulas. Let $\Omega \subset \mathbb{R}^3$ be a domain with boundary $\partial\Omega$. Use elementary integration by parts to derive the following integration by parts formulas:

$$\int_\Omega \boldsymbol{\nabla} u\, v = -\int_\Omega u\, \boldsymbol{\nabla} v + \int_{\partial\Omega} n u\, v\,,$$

$$\int_\Omega (\boldsymbol{\nabla} \times E) \cdot F = \int_\Omega E \cdot (\boldsymbol{\nabla} \times F) + \int_{\partial\Omega} (n \times E) \cdot F\,,$$

$$\int_\Omega (\boldsymbol{\nabla} \cdot u)\, v = -\int_\Omega u \cdot (\boldsymbol{\nabla} v) + \int_{\partial\Omega} u \cdot n\, v\,.$$

(3 points)

1.4.6. Maxwell problem. Repeat the discussion from Section 1.4.4 on the implicit satisfaction of Gauss's Magnetic Law and continuity equation for the variational formulation in terms of magnetic field H. (5 points)

Chapter 2

Coercivity

Coercive Problems

As we saw at the conclusion of Section 1.2, stability is a critical condition for convergence of the Galerkin method. In this chapter we study an important class of *coercive problems* for which the stability can be taken for granted. We begin by recalling an even more specialized class of coercive problems that originate from minimization of energy and discuss equivalence of the Galerkin method with the Ritz method. We study then the famous *Lax–Milgram Theorem* that uses the coercivity condition for the bilinear form to establish the well-posedness of the variational problem. We immediately link then the Lax–Milgram theory with *Cea's Lemma* to obtain the fundamental convergence result for coercive problems. In the concluding section, we revisit those of the earlier introduced model problems that satisfy the coercivity assumption and link the coercivity to ellipticity conditions.

2.1 ▪ Minimization Principle and the Ritz Method

Abstract minimization principle. The real case.   Assume the symmetric functional setting with trial and test spaces coinciding with each other, $U = V$. Assume additionally that the spaces are real, and consider bilinear and linear forms corresponding to the abstract variational formulation. Define the *quadratic energy functional (total potential energy)*,

$$J(u) := \frac{1}{2}b(u, u) - l(u) \,,$$

and derive the corresponding Gâteaux derivative,

$$\langle \delta J(u), v \rangle = \frac{1}{2}[b(u, v) + b(v, u)] - l(v) \,.$$

If we additionally assume that form b is *symmetric*, i.e.,

$$b(u, v) = b(v, u) \quad u, v \in U \,,$$

the formula reduces to

$$\langle \delta J(u), v \rangle = b(u, v) - l(v) \,.$$

The abstract variational formulation,

$$
\begin{cases}
u \in U \,, \\
b(u, v) = l(v) \quad v \in U \,,
\end{cases}
\tag{2.1.1}
$$

represents thus a necessary condition for u to be a minimizer (or maximizer as well).

Conversely, a simple computation reveals that

$$
J(u + v) - J(u) = b(u, v) - l(v) + \frac{1}{2} b(v, v) \,.
$$

If form $b(v, v)$ is *positive definite* over $U = V$, i.e.,

$$
b(v, v) > 0, \quad v \in V, \, v \neq 0 \,,
\tag{2.1.2}
$$

then solution u to the variational problem is seen to be the *unique minimizer* of the total potential energy functional $J(u)$.

The minimization problem

$$
u = \arg \min_{w \in U} J(w)
\tag{2.1.3}
$$

and the variational formulation (2.1.1) are thus *equivalent to each other*.

Well-posedness. Equivalence of the minimization and the variational problems does not prove that either of them is well-posed. The symmetry and positive definiteness of form $b(u, v)$ imply that $b(u, v)$ may be identified as an inner product with the corresponding *energy norm*

$$
\|u\|_E^2 = b(u, u) \,.
\tag{2.1.4}
$$

The well-posedness of the variational problem is implied then by the *Riesz Representation Theorem* [66], provided we can show that form $l(v)$ is continuous in the energy norm, and the space U equipped with the energy norm is complete. In order to guarantee these properties, we upgrade the assumption on positive definiteness of form $b(u, v)$ to the *coercivity condition*. We say that a sesquilinear and Hermitian form $b(u, v)$ is U-coercive if there exists a constant $\alpha > 0$ such that

$$
\alpha \|u\|_U^2 \leq b(u, u), \quad u \in U \,.
\tag{2.1.5}
$$

Note that the coercivity indeed implies positive definiteness. With the coercivity assumption in place, the original and energy norms are equivalent,

$$
\alpha \|u\|_U^2 \leq \|u\|_E^2 \leq M \|u\|_U^2 \,.
$$

Consequently, if $(U, \| \cdot \|_U)$ is complete then so is $(U, \| \cdot \|_E)$. By the same token, if $l(v)$ is continuous with respect to the norm $\| \cdot \|_U$, then it is also continuous with respect to the energy norm.

The complex case. All the considerations generalize to the case of a complex Hilbert space U and a coercive sesquilinear Hermitian form $b(u, v)$. The total potential energy functional is defined as

$$
J(u) := \frac{1}{2} b(u, u) - \Re l(u) \,.
$$

As form b is Hermitian, $b(u, u)$ is real but $l(u)$ is, in general, complex-valued, hence the necessity of using its real part only. The Gâteaux derivative of the energy functional is

$$
\langle \partial J(u), v \rangle = \Re(b(u, v) - l(v)) \,.
$$

The necessary condition for the minimizer (at first) is thus vanishing of the real part only. Recall, however, that for a linear (or antilinear) functional $l(v)$ defined on a complex space V, vanishing of the real part of the functional is equivalent to vanishing of the whole functional,

$$\Re l(v) = 0 \quad v \in V \qquad \Leftrightarrow \qquad l(v) = 0 \quad v \in V.$$

Indeed, let $l(v)$ be antilinear. Then

$$\Re l(iv) = \Re(-il(v)) = \Re(-i(\Re l(v) + i\Im l(v))) = \Im l(v).$$

The Ritz method. Assume $b(u, v)$ is Hermitian and U-coercive. Let $U_h \subset U$ be a finite-dimensional subspace of U. The following problems are equivalent to each other.

(i) Minimization of energy over the approximate space U_h:

$$J(u_h) = \min_{w_h \in U_h} J(w_h).$$

(ii) Galerkin approximation of the variational problem:

$$\begin{cases} u_h \in U_h, \\ b(u_h, v_h) = l(v_h) \quad \forall v_h \in U_h. \end{cases}$$

(iii) Minimization of distance between the exact and approximate solutions measured in the energy norm:

$$\|u - u_h\|_E = \min_{w_h \in U_h} \|u - w_h\|_E,$$

where $\|v\|_E^2 := b(v, v)$.

(iv) Minimization of the residual in the norm dual to the energy norm,

$$\|b(u_h, \cdot) - l(\cdot)\|_{U'} = \inf_{w_h \in U_h} \|b(w_h, \cdot) - l(\cdot)\|_{U'},$$

where

$$\|l\|_{U'} := \sup_{v \in U} \frac{|l(v)|}{\|v\|_E}.$$

Proof. Equivalence of (i) and (ii) has already been proved for space U. As U was an arbitrary inner product space, the result holds also for the finite-dimensional space U_h.

To see the equivalence of (i) and (iii), expand the formula for the energy norm,

$$\frac{1}{2}\|u - u_h\|_E^2 = \frac{1}{2}b(u - u_h, u - u_h) = \frac{1}{2}b(u, u) + \frac{1}{2}b(u_h, u_h) - \underbrace{\Re b(u, u_h)}_{= \Re l(u_h)} = \frac{1}{2}b(u, u) + J(u_h).$$

Equivalence with the fourth condition is left as an exercise; compare Exercise 2.1.4.   □

In terms of the energy norm, the Ritz method delivers the orthogonal projection (the best approximation error). In other words, if we equip the energy space with the energy norm, the Ritz method (equivalent to the Galerkin method) is stable with the stability constant equal to one.

Equivalence of the original and energy norms implies also stability of the discretization in the original norm. Indeed,

$$\alpha\|u - u_h\|_U^2 \leq \|u - u_h\|_E^2 = \inf_{w_h \in U_h} \|u - w_h\|_E^2 \leq M \inf_{w_h \in U_h} \|u - w_h\|_U^2,$$

which implies that

$$\|u - u_h\|_U \leq \underbrace{\sqrt{\frac{M}{\alpha}}}_{\text{stability constant}} \inf_{w_h \in U_h} \|u - w_h\|_U .$$

Exercises

2.1.1. Use the abstract minimization framework to identify energy functionals for the Poisson and elasticity problems. Verify positive definiteness of the corresponding bilinear forms. (3 points)

2.1.2. Consider the diffusion-reaction problem with $a_{ij} = \delta_{ij}$, $b_j = 0$, and $c > 0$ with *arbitrary* BCs. Identify the energy functional and verify positive definiteness of bilinear form. (3 points)

2.1.3. Consider again the diffusion-reaction problem discussed in Exercise 2.1.2 but with a relaxed condition for the reaction coefficient $c \geq 0$ (in particular, the reaction term may vanish) and the Cauchy BC imposed on the whole boundary Γ:

$$\frac{\partial u}{\partial n} + \beta u = g .$$

Derive the corresponding classical variational formulation and identify condition(s) for coefficient β for the bilinear form to be positive definite. (5 points)

2.1.4. Prove that the Ritz method is equivalent to the minimization of the residual measured in the norm dual to the energy norm. (3 points)

2.2 ▪ Lax–Milgram Theorem and Cea's Lemma

Coercive bilinear and sesquilinear forms.   We extend now the notion of coercivity to general, not necessarily Hermitian, sesquilinear forms. A sesquilinear form $b(u, v)$ defined on a Hilbert space U is *coercive* if there exists a positive (coercivity) constant $\alpha > 0$ such that

$$\Re b(u, u) \geq \alpha \|u\|_U^2 \quad \forall u \in U . \tag{2.2.6}$$

Condition (2.2.6) implies a weaker condition:

$$|b(u, u)| \geq \alpha \|u\|^2 . \tag{2.2.7}$$

The two conditions are not equivalent, e.g., sesquilinear form $b(x, y) := ix\bar{y}$, $x, y \in \mathbb{C}$, satisfies the condition above but $\Re b(u, u) = 0$, so the condition (2.2.6) is not satisfied. We shall use the stronger condition (2.2.6) in the elementary proof of the Lax–Milgram Theorem below but, otherwise, we will use condition (2.2.7). As we will learn in Chapter 4, both conditions imply the assumptions of Babuška–Nečas Theorem 4.1, and Cea's argument below, so the difference between the applicability of both conditions is a mere technicality.

Theorem 2.1 (Lax–Milgram Theorem). *Let U be a Hilbert space. Let $b(u, v)$ be a sesquilinear, continuous, and coercive form defined on $U \times U$. Let $l \in U'$. The (abstract) variational problem,*

$$\begin{cases} u \in U , \\ b(u, v) = l(v) \quad \forall v \in U , \end{cases}$$

is then well-posed, i.e., it admits a unique solution u that depends continuously upon the data, namely

$$\|u\|_U \le \frac{1}{\alpha}\|l\|_{U'}\,,$$

where α is the coercivity constant.

Proof. The Lax–Milgram Theorem is a corollary to the Babuška–Nečas Theorem, which in turn is a reformulation of the Banach Closed Range Theorem to variational problems. The following is an elementary proof reproduced from [10, p. 62], under the stronger version of coercivity condition (2.2.6). The proof relies on two theorems: the Riesz Representation Theorem and the Banach Contractive Map Theorem. Both of these results are considered to be more elementary than the Closed Range Theorem.

Consider the map

$$T_l u = u - \rho R^{-1}(Bu - l)\,,$$

where $B : U \to U'$ is the operator corresponding to bilinear form $b(u, v)$, and $R : U \to U'$ is the Riesz operator corresponding to the scalar product in U. We shall prove that, with a proper choice of constant $\rho > 0$, map $T_l : U \to U$ is a *contraction*, i.e., there exists a contraction constant $0 < k < 1$ such that

$$\|T_l u_1 - T_l u_2\|_U \le k\|u_1 - u_2\|_U\,.$$

By the Contractive Map Theorem, map T_l has then a unique fixed point u, i.e., $T_l u = u$, which is equivalent to $Bu = l$. The stability estimate follows directly from the coercivity assumption,

$$\alpha\|u\|_U^2 \le b(u,u) = |l(u)| \le \|l\|_{U'}\,\|u\|_U\,.$$

Notice that (affine) map T_l is a contraction iff linear map T_0 (i.e., with $l = 0$) is a contraction, i.e., $\|T_0\| < 1$.

We have now

$$\begin{aligned}
\|T_0 u\|_U^2 &= (u - \rho R^{-1}Bu, u - \rho R^{-1}Bu) \\
&= \|u\|_U^2 - \rho(R^{-1}Bu, u) - \rho(u, R^{-1}Bu) + \rho^2\|R^{-1}Bu\|_U \\
&= \|u\|_U^2 - \rho\langle Bu, u\rangle - \rho\overline{\langle Bu, u\rangle} + \rho^2\|R^{-1}Bu\|_U \\
&= \|u\|_U^2 - 2\rho\Re b(u,u) + \rho^2\|R^{-1}Bu\|_U \\
&\le \underbrace{(1 - 2\rho\alpha + \rho^2 M^2)}_{=k^2}\|u\|_U^2
\end{aligned}$$

since $\|R\| - \|R^{-1}\| - 1$, and $\|B\| \le M$. Selecting $\rho \in (0, 2\alpha/M^2)$, we get $k < 1$, which finishes the proof. □

Galerkin orthogonality. Let $U_h \subset U$ and $V_h \subset V$ be approximate trial and test spaces. Let $u_h \in U_h$ be the Galerkin approximation to the variational problem,

$$\begin{cases} u_h \in U_h\,, \\ b(u_h, v_h) = l(v_h) \quad \forall v_h \in V_h\,. \end{cases} \tag{2.2.8}$$

Testing the exact problem with approximate test functions,

$$b(u, v_h) = l(v_h) \quad \forall v_h \in V_h \subset V\,,$$

and subtracting the two equations from each other, we obtain the *Galerkin orthogonality* result:

$$b(u - u_h, v_h) = 0 \quad \forall v_h \in V_h \,. \tag{2.2.9}$$

Note that, in general, the form b may be neither Hermitian nor positive definite and, therefore, the orthogonality is not meant in the sense of a scalar product.

Theorem 2.2 (Cea's Lemma). *Let $b(u, v)$ be a continuous and coercive sesquilinear form defined on a Hilbert space U,*

$$|b(u, v)| \le M \|u\| \, \|v\|, \quad u, v \in U \,,$$

$$|b(v, v)| \ge \alpha \|v\|^2, \quad v \in U, \quad \alpha > 0 \,.$$

Let $U_h \subset U$, and let $u_h \in U_h$ be the Bubnov–Galerkin projection of some $u \in U$ onto subspace U_h, i.e.,

$$b(u - u_h, v_h) = 0 \quad \forall v_h \in U_h \,.$$

Then the following stability result holds:

$$\underbrace{\|u - u_h\|_U}_{\text{approximation error}} \le \frac{M}{\alpha} \underbrace{\inf_{w_h \in U_h} \|u - w_h\|_U}_{\text{the best approximation error}} \,. \tag{2.2.10}$$

Proof. We have

$$\alpha \|u - u_h\|_U^2 \le |b(u - u_h, u - u_h)| \qquad\qquad \text{(coercivity)}$$

$$= |b(u - u_h, u - w_h + w_h - u_h)|$$

$$= |b(u - u_h, u - w_h) + \underbrace{b(u - u_h, w_h - u_h)}_{= 0}| \qquad \text{(Galerkin orthogonality)}$$

$$\le M \|u - u_h\| \, \|u - w_h\|_U \qquad\qquad \text{(continuity)} \,,$$

which implies

$$\|u - u_h\| \le \frac{M}{\alpha} \inf_{w_h \in U_h} \|u - w_h\|_U \,. \qquad \square$$

Note that Cea's result does not provide an optimal stability constant for the Hermitian problems (compare with the Ritz method).

Exercises

2.2.1. Let U be a Hilbert space, and let $b(u, v)$ be a continuous, coercive form defined on $U \times U$. Let $U_h \subset U$ be a finite-dimensional subspace. Define the map

$$P_h : U \ni u \to u_h \in U_h \,,$$

where $u_h \subset U_h$ is the solution to the approximate variational problem:

$$\begin{cases} u_h \in U_h \,, \\ b(u_h, v_h) = b(u, v_h), \quad v_h \in U_h \,. \end{cases}$$

Show that map P_h is a well-defined, linear, and continuous projection, and estimate its norm. (3 points)

2.3 ▪ Examples of Problems Fitting the Ritz and Lax–Milgram and Cea Theories

2.3.1 ▪ A General Diffusion-Convection-Reaction Problem

We return to the classical diffusion-convection-reaction problem introduced in Section 1.3.

$$\begin{cases} -\dfrac{\partial}{\partial x_i}\left(a_{ij}\dfrac{\partial u}{\partial x_j}\right) + b_j\dfrac{\partial u}{\partial x_j} + cu = f & \text{in } \Omega\,, \\[2ex] \qquad\qquad\qquad\qquad\qquad u = u_0 & \text{on } \Gamma_1\,, \\[2ex] \qquad\qquad\qquad a_{ij}\dfrac{\partial u}{\partial x_j}n_i = g & \text{on } \Gamma_2\,, \\[2ex] \qquad\qquad a_{ij}\dfrac{\partial u}{\partial x_j}n_i + \beta u = g & \text{on } \Gamma_3\,. \end{cases} \qquad (2.3.11)$$

The *material data* consist of a symmetric diffusion tensor $a_{ij} = a_{ji}$, convection vector b_j, reaction coefficient c, and coefficient β present in the Cauchy (Robin) BC on Γ_3. The *load data* consist of functions f, u_0, g defined in Ω, Γ_1, and $\Gamma_2 \cup \Gamma_3$, respectively. The problem is real-valued. The bilinear and linear forms corresponding to the classical variational formulation are as follows:

$$b(u,v) = \int_\Omega \left\{ a_{ij}\frac{\partial u}{\partial x_j}\frac{\partial v}{\partial x_i} + b_j\frac{\partial u}{\partial x_j}v + c\,u\,v \right\} + \int_{\Gamma_3}\beta u v\,,$$
$$l(v) = \int_\Omega fv + \int_{\Gamma_2\cup\Gamma_3}gv\,. \qquad (2.3.12)$$

As discussed in Section 1.3, we assume that functions a_{ij}, b_j, c are bounded over $\overline{\Omega}$, and β is bounded over Γ_3,

$$\|a_{ij}(x)\| \le a_{\max} < \infty, \quad \|b_j(x)\| \le b_{\max} < \infty, \quad |c(x)| \le c_{\max} < \infty, \qquad x \in \overline{\Omega}\,.$$

In other words, they are L^∞-functions. Similarly, we assume

$$|\beta(x)| \le \beta_{\max} < \infty, \qquad x \in \Gamma_3\,.$$

The Cauchy–Schwarz inequality leads then to the choice of the energy spaces,

$$X = H^1(\Omega)\,,$$
$$V = \{v \in H^1(\Omega) \,:\, v = 0 \text{ on } \Gamma_1\}\,,$$
$$U = \{u \in H^1(\Omega) \,:\, u = u_0 \text{ on } \Gamma_1\} = \tilde{u}_0 + V\,,$$

where $\tilde{u}_0 \in X$ is a finite energy lift of Dirichlet data u_0. Boundary values are understood in the sense of the *Trace Theorem*, or simply *in the sense of traces*. This implies a regularity assumption on the Dirichlet data $u_0 \in H^{1/2}(\Gamma_1)$. A continuous function u_0 will do but a discontinuous one *will not*. In order to ensure the continuity of linear form l, we may assume $f \in L^2(\Omega)$ and $g \in L^2(\Gamma_2 \cup \Gamma_3)$. As discussed in Section 1.2, the nonhomogeneous Dirichlet data is accounted for by representing $u = \tilde{u}_0 + w$, $w \in V$ and solving for w,

$$\begin{cases} w \in V\,, \\ b(w,v) = l_{\mathrm{mod}}(v), \quad v \in V\,, \end{cases} \qquad (2.3.13)$$

where l_{mod} is the *modified linear form*,

$$l_{\mathrm{mod}}(v) := l(v) - b(\tilde{u}_0, v).$$

With the regularity assumptions made so far, both bilinear and linear forms are continuous.

Our first observation concerns symmetry of form b, i.e., necessary and sufficient conditions for $b(u, v) = b(v, u)$. It is easy to see that both diffusion and reaction terms are symmetric. In the case of the diffusion term, this is a consequence of the symmetry of the diffusion tensor, $a_{ij} = a_{ji}$. It is equally easy to see that the convection term can never be symmetric. Hence our first observation: in the presence of convection, Ritz theory is not applicable.

We shall look now for possible assumptions to secure coercivity of bilinear form $b(u, v)$. The problem is said to be *elliptic* if the diffusion tensor is positive definite. More precisely,

$$a_{ij}\xi_i\xi_j \geq 0 \quad \forall \xi_i \quad \text{and} \quad a_{ij}\xi_i\xi_j = 0 \Rightarrow \xi_i = 0.$$

A symmetric $N \times N$ tensor has N real eigenvalues, and the relation above translates into the assumption that all N eigenvalues are positive; compare Exercise 2.3.1. As the tensor changes with x, its smallest eigenvalue depends also upon x, $\lambda_{\min} = \lambda_{\min}(x)$. We make a stronger assumption that $\lambda_{\min}(x)$ is bounded away from zero,

$$\lambda_{\min}(x) \geq a_0 > 0, \quad x \in \overline{\Omega}. \tag{2.3.14}$$

This is equivalent (see again Exercise 2.3.1) to the assumption

$$a_{ij}(x)\,\xi_i\xi_j \geq a_0\,\xi_i\xi_i, \quad x \in \overline{\Omega}. \tag{2.3.15}$$

We say that the problem is *uniformly* or *strictly elliptic*. With the uniform ellipticity assumption, the diffusion term in the bilinear term is bounded below by the H^1-seminorm,

$$\int_\Omega a_{ij}\frac{\partial u}{\partial x_j}\frac{\partial u}{\partial x_i} \geq a_0|u|^2_{H^1(\Omega)}.$$

This is *almost* the coercivity condition. The L^2-part of the H^1-norm can be controlled in many ways. The most common one is through the essential BC on Γ_1.

Lemma 2.3 (Poincaré Inequality). *Let Ω be a bounded domain in $\mathbb{R}^N$, and let Γ_1 have a positive measure. There exists a positive constant $\alpha > 0$ such that*

$$\alpha\|v\|^2 \leq \|\boldsymbol{\nabla}v\|^2, \quad v \in V := \{v \in H^1(\Omega) : v = 0 \text{ on } \Gamma_1\}. \tag{2.3.16}$$

Proof. We obtain the proof by contradiction. Suppose there exists a sequence $v_n \in V$ such that

$$\|v_n\| = 1 \quad \text{and} \quad \|\boldsymbol{\nabla}v_n\| \to 0.$$

From every bounded sequence in a Hilbert space, we can extract a weakly convergent subsequence (denoted with the same symbol) $v_n \rightharpoonup v_0 \in V$. Weak convergence of $v_n \rightharpoonup v_0$ in $H^1(\Omega)$ implies weak convergence of $\boldsymbol{\nabla}v_n \rightharpoonup \boldsymbol{\nabla}v_0$ in $L^2(\Omega)$. By the lower weak sequential semicontinuity of the L^2-norm, $\boldsymbol{\nabla}v_0 = 0$, i.e., v_0 must be a constant. BC on Γ_1 implies that v_0 must vanish and, therefore, $\|v_0\| = 0$. By the Rellich Theorem (see [27, Theorem 3.7.2]), sequence $v_n \to v_0$ in $L^2(\Omega)$. But the convergence in L^2-norm implies that $\|v_0\| = 1$, a contradiction. $\qquad\square$

The proof is very standard. For an example of a more constructive and elementary proof, see Exercise 2.3.5. The Poincaré inequality implies now immediately the coercivity condition for the diffusion part. We have

$$\begin{array}{rcl} \|\nabla v\|^2 & = & \|\nabla v\|^2 \\ \|v\|^2 & \leq & \alpha^{-1}\|\nabla v\|^2 \\ \hline \|v\|^2_{H^1(\Omega)} & \leq & (1+\alpha^{-1})\|\nabla v\|^2 \end{array}$$

This implies

$$a_0(1+\alpha^{-1})^{-1}\|v\|^2_{H^1(\Omega)} \leq a_0\|\nabla v\|^2 \leq \int_\Omega a_{ij}\frac{\partial v}{\partial x_j}\frac{\partial v}{\partial x_i}\,.$$

The convection and reaction terms may help, stay neutral, or disturb the coercivity condition. Of course, if they vanish, i.e., we have a pure diffusion problem only, we are done. If the reaction coefficient is nonnegative $c \geq 0$, the corresponding reaction contribution is nonnegative as well,

$$\int_\Omega c v^2 \geq 0\,,$$

and the combined diffusion plus reaction term represents a coercive form. If the reaction term is uniformly bounded away from zero,

$$c(x) \geq c_0 > 0\,, x \in \overline{\Omega}\,,$$

we have

$$c_0\|v\|^2 \leq \int_\Omega c v^2\,.$$

In this case, we can claim coercivity over the whole H^1-space, i.e., without the help of the Dirichlet BC and the Poincaré inequality,

$$\min\{a_0, c_0\}\|v\|_{H^1(\Omega)} \leq c_0\|v\|^2 + a_0\|\nabla v\|^2 \leq \int_\Omega a_{ij}\frac{\partial v}{\partial x_j}\frac{\partial v}{\partial x_i} + c v^2\,, \quad v \in H^1(\Omega)\,.$$

If reaction coefficient c is negative, the situation is not entirely hopeless, provided the coefficient is not too large. More precisely, if

$$|c(x)| < a_0^{-1}(1+\alpha^{-1})\,,$$

then the sum of diffusion and reaction terms is still coercive.

The same comment applies to the convective term. If the problem is *diffusion dominated*, the sum of the diffusion and convection terms may be coercive. More precisely, the continuity estimate,

$$\left|\int_\Omega b_j \frac{\partial v}{\partial x_j} v\right| \leq b_{\max}\|v\|^2_{H^1(\Omega)}\,,$$

implies that

$$-b_{\max}\|v\|^2_{H^1(\Omega)} \leq \int_\Omega b_j \frac{\partial v}{\partial x_j} v\,.$$

Thus, if

$$a_0(\alpha^{-1}+1)^{-1} - b_{\max} > 0\,,$$

the sum of the diffusion and convective term will represent a V-coercive bilinear form. It is less intuitive to see that, with the appropriate assumptions, the convective term may not disturb coercivity at all or even help it. We have

$$\int_\Omega b_j \frac{\partial u}{\partial x_j} u = \int_\Omega b_j \frac{\partial}{\partial x_j}\left(\frac{1}{2}u^2\right) = \frac{1}{2}\int_\Omega (-\operatorname{div} b)u^2 + \frac{1}{2}\int_{\Gamma_2\cup\Gamma_3} b_j n_j\, u^2\,.$$

If div $b \leq 0$, the first term is nonnegative. In the particular, important case of an incompressible advection, div $b = 0$, the term vanishes. If parts Γ_2 and Γ_3 of the boundary are contained in the *outflow boundary*,

$$\Gamma_{\text{out}} := \{x \in \Gamma \,:\, b_n(x) = b_j(x)n_j \geq 0\}\,,$$

then the second term is also nonnegative.

Note finally that, with $\beta \geq 0$, the boundary contribution to the bilinear form stays nonnegative as well. As you can see, it makes little sense to attempt to formulate various scenarios guaranteeing coercivity of the bilinear form b. It is a skill that needs to be acquired to check (see) whether a particular bilinear form is coercive. In the end, it is an interplay of the elements we have used above: strict ellipticity, Poincaré inequality, integration by parts, and appropriate assumptions on BCs and material data; compare Exercises 2.3.2 and 2.3.3.

Remark 2.4. In the case of the nonhomogeneous Dirichlet condition, the modified linear functional depends upon the lift of BC data, i.e., upon the way we extend u_0 into the domain. Consequently, solution w to the (modified) problem with homogeneous Dirichlet condition will depend upon the lift as well, and so will the ultimate solution u. In the FE practice we proceed in a different way. We first interpolate (project) boundary data u_0 into the trace of FE space $X_h \subset X$, replacing u_0 with some approximation $u_{0,h}$. Then we use the FE basis (shape) functions to lift the approximate BC data $u_{0,h}$ into the FE space to obtain $\tilde{u}_{0,h}$. FE approximation $w_h \in V_h \subset V$ still does depend upon the way we lift the approximate data but the ultimate FE solution $u_h = w_h + \tilde{u}_{0,h}$ *does not*. This is the good news. The bad news is that, in the error analysis (and control), we have to account now for the error in approximating the Dirichlet data. We are simply solving a "wrong problem." Most research papers ignore this error by assuming that your original Dirichlet data live in the trace of the FE space. In many cases (polynomial data), this condition is indeed satisfied.

2.3.2 ▪ Linear Elasticity

We turn now to the second classical example introduced in Section 1.3.2—the linear elastostatics problem. In the following discussion, we will restrict ourselves to the case of kinematic and traction BCs only.

$$\begin{cases} -\sigma_{ij,j} = f_i & \text{in } \Omega\,, \\ \quad u_i = u_{0,i} \text{ on } \Gamma_1\,, \\ \quad t_i = g_i & \text{on } \Gamma_2\,, \end{cases}$$

where the stresses σ_{ij} and tractions t_i are functions of displacements u_i,

$$\sigma_{ij} = E_{ijkl}\epsilon_{kl} = E_{ijkl}\,u_{k,l}\,,$$

$$t_i = \sigma_{ij}n_j = E_{ijkl}\,u_{k,l}n_j\,.$$

The material data are represented by elasticities E_{ijkl}, whereas the load data include volume body force f_i, prescribed displacements $u_{0,i}$ on Γ_1, and prescribed tractions g_i on Γ_2. The elasticity tensor satisfies the following conditions.

$$E_{ijkl} = E_{jikl} = E_{ijlk} \qquad \text{(minor symmetries)}\,,$$

$$E_{ijkl} = E_{klij} \qquad \text{(major symmetry)}\,,$$

$$E_{ijkl}\xi_{ij}\xi_{kl} \geq a_0\xi_{ij}\xi_{ij} \qquad \forall \xi_{ij} = \xi_{ji}\,, a_0 > 0\,.$$

As in the definition of strictly elliptic problems, the last condition represents an upgrade of the condition on positive definiteness of tensor of elasticities. Note that, by definition, elasticities represent a positive definite operator acting on symmetric 2-tensors (and symmetric only). In the case of an isotropic material,

$$E_{ijkl} = \mu(\delta_{ik}\delta_{jl} + \delta_{il}\delta_{jk}) + \lambda\delta_{ij}\delta_{kl} \,,$$

where $\mu, \lambda > 0$ are Lamé constants. The formulas for the bilinear and linear forms corresponding to the classical variational formulation (Principle of Virtual Work) are as follows:

$$b(u,v) := \int_{\Omega} E_{ijkl}u_{k,l}v_{i,j} \,,$$

$$l(v) := \int_{\Omega} f_i v_i + \int_{\Gamma_2} g_i v_i \,.$$

The Cauchy–Schwarz inequality leads to the choice of the energy spaces:

$$X = (H^1(\Omega))^N \,,$$
$$V = \{v \in X : v_i = 0 \text{ on } \Gamma_1\} \,,$$
$$U = \{u \in X : u_i = u_{0,i} \text{ on } \Gamma_1\} = \tilde{u}_0 + V \,,$$

where $\tilde{u}_0 \in X$ is a finite energy lift of u_0.

At first, we are tempted to reproduce the reasoning from the analysis of the diffusion problem and use the strict ellipticity condition to claim boundedness below with the H^1-seminorm. We cannot do it though since the ellipticity condition is satisfied only for symmetric tensors ξ_{ij}. In other words, we control only the symmetric part of the displacement gradient,

$$\int_{\Omega} E_{ijkl}u_{k,l}u_{i,j} = \int_{\Omega} E_{ijkl}\epsilon_{ij}\epsilon_{kl} \geq a_0 \int_{\Omega} \epsilon_{ij}\epsilon_{ij} = \sum_{ij}\|\epsilon_{ij}\|^2 \,.$$

This is where the fundamental result of Korn comes in.

Theorem 2.5 (Korn's Inequality [55]). *Let Ω be a bounded Lipschitz domain in $\mathbb{R}^N$. There exists a positive constant $C_K > 0$ such that*

$$C_K\|u\|_{H^1(\Omega)}^2 \leq \|u\|_{L^2(\Omega)}^2 + \sum_{i,j}\|\epsilon_{ij}(u)\|_{L^2(\Omega)}^2 \quad \forall u \in (H^1(\Omega))^N \,, \tag{2.3.17}$$

where $\epsilon_{ij}(u) = \frac{1}{2}(u_{i,j} + u_{j,i})$ is the symmetric part of ∇u (linearized strain). Constant C_K depends upon the domain but it is independent of u.

With the help of Korn's inequality and the kinematic BC on Γ_1, we can prove now that the strain energy controls the L^2-norm.

Theorem 2.6. *Let the assumptions of Korn's inequality hold. Let Γ_1 be a subset of boundary $\partial\Omega$ with nonzero measure. Then there exists constant $a_1 > 0$ such that*

$$a_1\|v\|_{L^2(\Omega)}^2 \leq \sum_{i,j}\|\epsilon_{ij}(v)\|_{L^2(\Omega)}^2 \quad \forall v \in (H^1(\Omega))^N : v = 0 \text{ on } \Gamma_1 \,. \tag{2.3.18}$$

Proof. We proceed by contradiction. Let v_n be a sequence such that $\|v_n\|_{L^2(\Omega)} = 1$, and the right-hand side above converges to zero. By Korn's inequality, sequence v_n is bounded in $H^1(\Omega)$. Consequently, we can extract from v_n a subsequence, denoted with the same symbol, converging weakly to a limit v, $v_n \rightharpoonup v$ in $H^1(\Omega)$. Next we observe that the L^2-norm of the strain is positive definite. Indeed, if it vanishes, v must be a rigid body motion and the kinematic BC sets it to zero. Positive definiteness implies strict convexity. In turn, strict convexity and (strong) continuity imply weak lower semicontinuity. Consequently,

$$\sum_{i,j} \|\epsilon_{ij}(v)\|^2_{L^2(\Omega)} \leq \liminf_{n \to \infty} \sum_{i,j} \|\epsilon_{ij}(v_n)\|^2_{L^2(\Omega)} = 0$$

and, therefore, the weak limit must also be a rigid body motion. The kinematic BC implies then again that $v = 0$. Finally, by the Rellich Embedding Theorem, $v_n \to 0$ in the L^2-norm. This is a contradiction with the assumption that $\|v_n\|_{L^2(\Omega)} = 1$ (the limit should have a unit L^2-norm as well).   □

We can now pull all the results together to estimate the coercivity constant,

$$\alpha \geq a_0(1 + a_1^{-1})^{-1} C_K \,.$$

Finally, note that the bilinear form is symmetric, which means that the Ritz theory applies.

If we switch from elastostatics to time-harmonic elastodynamics, we arrive at complex-value functions. The new sesquilinear form $b(u,v)$ includes an extra contribution corresponding to the inertia,

$$b(u,v) = \int_\Omega E_{ijkl} u_{k,l} \bar{v}_{i,j} - \omega^2 \int_\Omega \rho\, u_i \bar{v}_i \,,$$

where ρ is the density and ω denotes the angular velocity. The zero order term corresponds to the reaction term in the diffusion-reaction problem and, similarly to the case there, Hermitian form $b(u,v)$ has a chance to be coercive, provided frequency ω is sufficiently small. For a general ω, however, sesquilinear form $b(u,v)$ is no longer coercive so neither the Ritz nor the Lax–Milgram and Cea theories apply. We will study this class of problems in Section 4.2.

2.3.3 ▪ Model Curl-Curl and Grad-Div Problems

The following projection problem is encountered after time discretization of Maxwell transient problems:

$$\begin{cases} E \in H(\text{curl}, \Omega), \ n \times E = 0 \text{ on } \Gamma_1 \,, \\[2mm] \displaystyle\int_\Omega \boldsymbol{\nabla} \times E \cdot \boldsymbol{\nabla} \times \bar{F} + \epsilon \int_\Omega E \cdot \bar{F} = \int_\Omega f \cdot \bar{F} + \int_{\Gamma_2} g \cdot \bar{F}_t \,, \qquad (2.3.19) \\[2mm] F \in H(\text{curl}, \Omega), \ n \times F = 0 \text{ on } \Gamma_1 \,, \end{cases}$$

where F_t is the tangential component of F on boundary Γ, $F_t := -n \times (n \times F) = F - (F \cdot n)n$. Note that $g \cdot F_t = g_t \cdot F_t$ so g is assumed to be tangent to boundary Γ.

We start with the *Helmholtz decomposition* of E. Given $E \in H(\text{curl}, \Omega)$, $n \times E = 0$ on Γ_1, we seek

$$\begin{cases} p \in H^1(\Omega), \ p = 0 \text{ on } \Gamma_1 \,, \\[2mm] (\boldsymbol{\nabla} p, \boldsymbol{\nabla} q) = (E, \boldsymbol{\nabla} q), \quad q \in H^1(\Omega), \ q = 0 \text{ on } \Gamma_1 \,. \end{cases}$$

Obviously, p is well-defined. The decomposition

$$E = \underbrace{E - \boldsymbol{\nabla} p}_{=:E_0} + \boldsymbol{\nabla} p$$

is known as the *Helmholtz decomposition* of E. Note that, by construction,

$$E_0 \in V := \{E \in H(\text{curl}, \Omega) \ : \ n \times E = 0 \text{ on } \Gamma_1 \text{ and } (E, \nabla q) = 0 \quad \forall q \in H^1(\Omega), \ q = 0 \text{ on } \Gamma_1\}.$$

The next result is an analogue of the Poincaré inequality for $H(\text{curl}, \Omega)$ space.

Lemma 2.7 (Friedrichs' Inequality).   *Let Ω be a bounded domain in $\mathbb{R}^3$, and Γ_1 a part of boundary Γ with nonzero measure. There exists then a $C_F > 0$ such that*

$$C_F \|E\| \leq \|\nabla \times E\|, \qquad E \in V. \tag{2.3.20}$$

Proof. We present a proof for a simply connected domain Ω. The crucial argument in the proof is the compact embedding of space V into $L^2(\Omega)$ [70].

Assume, to the contrary, that there exists a sequence $E_n \in V$ such that

$$\|E_n\| = 1 \quad \text{and} \quad \|\nabla \times E_n\| \to 0,$$

in particular, E_n is bounded in V. As V is Hilbert, there exists a subsequence, denoted with the same symbol, converging weakly to a function $E \in V$. The weak lower semicontinuity of the norm implies

$$\|\nabla \times E\| \leq \liminf_{n \to \infty} \|\nabla \times E_n\| = 0.$$

Consequently, $E = \nabla p, \ p \in H^1(\Omega), p = 0$ on Γ_1. But

$$(E, \nabla p) = \|\nabla p\|^2 = 0 \quad \Rightarrow \quad \nabla p = E = 0.$$

At the same time, due to the compact embedding of V into $L^2(\Omega)$, $E_n \to E$ strongly in $L^2(\Omega)$, which implies that $\|E\| = 1$, a contradiction.   □

With $\epsilon > 0$, the problem is obviously coercive although the coercivity constant depends upon ϵ. And yet, with appropriate assumptions on the load, the solution may be bounded *uniformly* in ϵ.

Consider problem (2.3.19) and assume that domain Ω is simply connected. Let $E = E_0 + \nabla p$ be the Helmholtz decomposition of E. If the gradient part is missing, $\nabla p = 0$, Friedrichs' inequality implies that the L^2-norm of E is controlled by the L^2-norm of the curl. In order to eliminate the gradients from the solution, we need to assume that the load is orthogonal to the gradients, i.e.,

$$\int_\Omega f \cdot \overline{\nabla q} + \int_{\Gamma_2} g \cdot \overline{\nabla q}_t = 0, \quad q \in H^1(\Omega), q = 0 \text{ on } \Gamma_1.$$

Then, testing both sides of (2.3.19) with $F = \nabla p$, we obtain

$$\epsilon(E, \nabla p) = \epsilon \|\nabla p\|^2 = 0 \quad \Rightarrow \quad p = 0.$$

Finally, testing in (2.3.19) with $F = E_0$, we get

$$(1 + C_F^{-1})^{-1} \|E_0\|^2_{H(\text{curl}, \Omega)} \leq \|\nabla \times E_0\| \qquad \text{(Friedrichs' inequality)}$$

$$\leq \|\nabla \times E_0\|^2 + \epsilon \|E_0\|^2$$

$$\leq \|f\| \, \|E_0\| + \|g\|_* \, \|E_{0,t}\|_{H^{-1/2}(\text{curl}_\Gamma, \Gamma_2)}$$

$$\leq (\|f\| + C\|g\|_*)\|E_0\|_{H(\text{curl}, \Omega)},$$

which results in the ϵ-independent bound,

$$\|E_0\| \leq (1 + C_F^{-1})(\|f\| + C\|g\|_*)\,.$$

Above, C denotes the continuity constant of the tangential trace operator [27]

$$\gamma_t \,:\, H(\mathrm{curl}, \Omega) \ni E \to E_t \in H^{-1,2}(\mathrm{curl}_\Gamma, \Gamma)\,.$$

Here $H^{-1,2}(\mathrm{curl}_\Gamma, \Gamma)$ denotes the trace space,

$$H^{-1,2}(\mathrm{curl}_\Gamma, \Gamma) := \{E \in H^{-1/2}(\Gamma) \,:\, \mathrm{curl}_\Gamma\, E \in H^{-1/2}(\Gamma)\}\,,$$

and the star in $\|g\|_*$ denotes the dual norm to the norm in the space of restrictions of functions from $H^{-1,2}(\mathrm{curl}_\Gamma, \Gamma)$ to the Γ_2 part of the boundary. These are very nontrivial and rather technical details concerning energy space $H(\mathrm{curl}, \Omega)$.

Similar results hold for a model problem encountered after time-discretization of acoustics equations formulated in terms of velocity,

$$\begin{cases} u \in H(\mathrm{div}, \Omega),\ u_n = 0 \text{ on } \Gamma_1\,, \\[2mm] \displaystyle\int_\Omega \boldsymbol{\nabla} \cdot u\, \boldsymbol{\nabla} \cdot v + \epsilon \int_\Omega u \cdot v = \int_\Omega f \cdot v + \int_{\Gamma_2} g v_n\,, \\[2mm] v \in H(\mathrm{div}, \Omega),\ v_n = 0 \text{ on } \Gamma_1\,. \end{cases} \qquad (2.3.21)$$

Lemma 2.8 (Friedrichs' Inequality for $H(\mathrm{div})$ Spaces). *Let Ω be a bounded domain in $\mathbb{R}^3$, and Γ_1 a part of boundary Γ with nonzero measure. Define the space*

$$V := \{v \in H(\mathrm{div}, \Omega) \,:\, v \cdot n = 0 \text{ on } \Gamma_1 \text{ and } (v, \boldsymbol{\nabla} \times F) = 0$$

$$\forall F \in H(\mathrm{curl}, \Omega),\ n \times F = 0 \text{ on } \Gamma_1\}\,.$$

There then exists a $C > 0$ such that

$$C\|v\| \leq \|\boldsymbol{\nabla} \cdot v\|, \qquad v \in V\,. \qquad (2.3.22)$$

Proof. The proof is fully analogous to the proof of Lemma 2.7.   □

Exercises

2.3.1. Let a_{ij} be a Hermitian tensor. Prove that a is positive definite iff all eigenvalues of the matrix are positive. Then prove that conditions (2.3.14) and (2.3.15) are equivalent. Does it make sense to speak about positive definiteness for non-Hermitian matrices? (2 points)

2.3.2. Coercivity. Consider the diffusion-convection-reaction model problem. Prove or disprove that the bilinear forms corresponding to the following data are V-coercive.

  (i) $b = c = 0$, $\mathrm{meas}\,\Gamma_3 > 0$, $\beta > 0$. *Hint:* Prove a slightly different version of the Poincaré inequality.

   Lemma 2.9. *Let $\Omega \subset \mathbb{R}^N$ be a bounded domain. There exists then a positive constant $\alpha > 0$ such that*

$$\alpha\|v\|^2 \leq \|\boldsymbol{\nabla} v\|^2 + \phi(v), \qquad v \in H^1(\Omega)\,,$$

where $\phi(v) = c(v,v)$ where $c(u,v)$ is a continuous, semipositive definite bilinear form defined on $H^1(\Omega)$ such that

$$\phi(\text{const}) = 0 \quad \Rightarrow \quad \text{const} = 0.$$

Then identify the appropriate functional ϕ.

(ii) $b = 0, c = -\omega^2$ with a small (frequency) ω, meas $\Gamma_1 > 0$.

(iii) $b = 0, c \geq 0, c \geq c_0 > 0$ on $\Omega_0 \subset \Omega$ with meas $\Omega_0 > 0$.

(5 points)

2.3.3. **Nonlocal terms.** Prove that the following bilinear form is coercive over the whole $H^1(\Omega)$ space.

$$b(u,v) := \int_\Omega \boldsymbol{\nabla} u \boldsymbol{\nabla} v + \int_\Omega u \int_\Omega v.$$

(5 points)

2.3.4. **Distributional derivatives** (compare Exercise 1.4.5). Let a domain $\Omega \subset \mathbb{R}^N$, $N = 2, 3$, be split into two subdomains Ω_1, Ω_2 with a smooth interface Γ. Let u, E, v be functions consisting of two smooth branches u^I, E^I, v^I, $I = 1, 2$, defined in the subdomains. By "smooth" we understand $u^I \in C^1(\overline{\Omega_I})$, etc. Let n be the unit vector on interface Γ pointing from subdomain Ω_1 into subdomain Ω_2.

(i) Let $\phi \in C_0^\infty(\Omega)$ be a Schwartz test function (scalar- or vector-valued). Use elementary integration by parts to derive the following formulas:

$$-\int_\Omega u \boldsymbol{\nabla} \phi = \sum_I \int_{\Omega_I} \boldsymbol{\nabla} u^I \phi \quad + \int_\Gamma [u] n \phi,$$

$$\int_\Omega E \boldsymbol{\nabla} \times \phi = \sum_I \int_{\Omega_I} \boldsymbol{\nabla} \times E^I \phi + \int_\Gamma [n \times E] \phi,$$

$$-\int_\Omega v \boldsymbol{\nabla} \cdot \phi = \sum_I \int_{\Omega_I} \boldsymbol{\nabla} \cdot v^I \phi \quad + \int_\Gamma [v \cdot n] \phi,$$

where

$$[u] = u^2 - u^1, \quad [n \times E] = n \times (E^2 - E^1), \quad [v \cdot n] = (v^2 - v^1) \cdot n.$$

(ii) Interpret the formulas above in the language of distributions using the definition of regular distributions, distributional derivatives, and corresponding operators of grad, curl, and div understood in the distributional sense. You will have to introduce a multidimensional equivalent of Dirac's delta.

(iii) Conclude that functions u, E, v belong to energy spaces $H^1(\Omega)$, $H(\text{curl}, \Omega)$, $H(\text{div}, \Omega)$ iff the corresponding continuity conditions across the interface Γ are satisfied:

$$[u] = 0, \quad [n \times E] = 0, \quad [v \cdot n] = 0.$$

(5 points)

2.3.5. **Elementary proof of the Poincaré inequality.**

(i) Use elementary means to prove the 1D version of the Poincaré inequality:

$$\alpha \int_0^1 |u|^2 \leq \int_0^1 |u'|^2 \quad \forall u \in H^1(0,1) : u(0) = 0, \quad \alpha > 0.$$

Provide a concrete estimate for α. *Hint:* Apply the Second Fundamental Theorem of Differential Calculus to interval $(0, x)$,

$$u(x) = \int_0^x u'(s)\, ds \,,$$

and take it from there.

(ii) Interpret the best (largest) Poincaré constant α as the minimum eigenvalue of the 1D Laplace operator with appropriate BC. Use the Sturm–Liouville Theorem to compute α and compare it with the estimate obtained in the previous step.

(iii) Use scaling arguments to derive the best Poincaré constant for an interval of length l to see how α changes with the size of the domain.

(iv) Repeat the first three steps for an elementary 2D scenario with $\Omega = (0,1)^2$ and u vanishing on the west boundary: $u(0, y) = 0, y \in (0, 1)$ (you will need to refresh your skills in separation of variables).

(5 points)

2.3.6. **Linearized rigid body motion.** Displacement $u = \omega \times x + a$, where $\omega, a \in \mathbb{R}^3$, is called a *linearized rigid body motion* with a representing a *translation* and ω an *infinitesimal rotation vector*. Prove that $\epsilon_{ij}(u) = 0$ iff u is a linearized rigid body motion. (3 points)

2.3.7. **Coercivity of elasticity bilinear form.** Application of Korn's inequality requires control of the L^2-norm of displacement u. In the text we have turned things around and have shown how the Korn inequality and kinematic BCs imply control of $\|u\|$. In this exercise, we seek more direct and elementary arguments to control the L^2-norm directly with the elastic energy $b(u, u)$.

(i) Consider the elasticity problem in a square domain $(0, 1)^2$ with kinematic BC on the south and west boundaries,

$$u(x_1, 0) = 0, \ x_1 \in (0, 1) \quad \text{and} \quad u(0, x_2) = 0, \ x_2 \in (0, 1) \,.$$

Use elementary means similar to those in Exercise 2.3.5 to prove that there exists a positive constant $C > 0$ such that

$$C \int_\Omega |v|^2 \le \int_\Omega \sum_{ij} |\epsilon_{ij}(v)|^2 \quad \text{for every kinematically admissible } v \in (H^1(\Omega))^2 \,.$$

(ii) Use the standard assumptions on the elasticities to conclude that the elastic bilinear form,

$$b(u, v) = \int_\Omega E_{ijkl} u_{k,l} v_{i,j} \,,$$

bounds the L^2-norm of kinematically admissible displacements.

(iii) Interpret the best (largest) L^2 boundedness below constant as the smallest elastic eigenfrequency (with density $\rho = 1$),

$$\begin{cases} u \in V_0, \ \lambda \in \mathbb{R} \,, \\ b(u, v) = \lambda(u, v) \quad \forall v \in V_0 \,, \end{cases}$$

where V_0 is the space of kinematically admissible displacements. Use a scaling argument to estimate $\alpha = \lambda_{\min}$ in terms of the size of the domain.

(5 points)

2.3.8. **Effect of BCs on coercivity of elastic bilinear form.** Consider the elastostatics problem with more complicated BCs imposed on a part of the boundary (with nonzero measure),

Case 1: $\quad u_n = 0, \quad t_t = g,$

Case 2: $\quad u_t = 0, \quad t_n = g,$

Case 3: $\quad t_i = \beta_{ij} u_j,$

Case 4: $\quad t_n = \beta u_n, \quad t_t = g.$

Here u_n, u_t and t_n, t_t are the normal and tangential components of displacement u_i or traction t_i, respectively, $\beta > 0$, and

$$\beta_{ij}\xi_i\xi_j \geq \beta_0\xi_i\xi_j, \quad \beta_0 > 0.$$

Discuss the effect of various BCs on coercivity of the bilinear form. Does it depend upon the shape of the domain? Discuss the case of a square versus a circular domain $\Omega \subset \mathbb{R}^2$. (5 points)

Chapter 3

Conforming Elements and Interpolation Theory

In this chapter we discuss the construction of finite elements corresponding to the exact grad-curl-div sequence energy spaces. The exposition is not intended to replace a systematic construction of various finite elements available in the literature, starting with Ciarlet's classic [18] and ending with Doug Arnold's *The Periodic Table of the Finite Elements* [1, 2]; see also https://www-users.cse.umn.edu/~arnold/femtable/. Instead, we try to communicate the main logic behind the construction of various H^1-, $H(\mathrm{curl})$-, $H(\mathrm{div})$-, and L^2-conforming elements and illuminate the difference between Ciarlet's construction of interpolation operators and the *projection-based (PB) interpolation*.

3.1 ▪ H^1-Conforming Finite Elements

3.1.1 ▪ Classical H^1-Conforming Elements

Courant's triangle. The FE method is a special case of the Galerkin method where the basis functions are constructed by "gluing" together polynomials defined on individual elements. First, domain $\Omega \subset \mathbb{R}^N$ is covered with a finite element mesh consisting of elements K. Next, we define our FE discretization by defining element *shape functions* $\phi_j = \phi_{j,K}$ defined on individual elements K, and, finally, we glue the element shape functions into global Galerkin *basis functions* e_i. Note the terminology: shape functions are defined on a single element K, basis functions are defined on domain Ω.

By the results discussed in Exercise 2.3.4, a function is H^1-conforming, i.e., it lives in the energy space $H^1(\Omega)$, iff it is *globally continuous*. The basis functions need to be globally continuous.

The first and perhaps the simplest construction came from Richard Courant for the case of a polygonal domain $\Omega \subset \mathbb{R}^2$ covered with a regular triangular mesh.[5] For each vertex node v_i in the mesh, Courant constructed a basis function e_i that assumes value one at v_i, is zero at the remaining vertex nodes, and, over each element K, is a linear polynomial. The concept is illustrated in Fig. 3.1. As we explode the function into the adjacent element contributions, we see that the function is a union of linear vertex shape functions from elements adjacent to the vertex, extended by zero to the rest of the mesh. Each triangular element comes with three vertex shape functions. The approximate solution in element K is constructed as a linear combination of

---

[5]A mesh is said to be *regular* if every vertex node in the mesh constitutes also a vertex for each adjacent triangle. See [25] for examples of irregular meshes with *hanging nodes*.

Figure 3.1: Courant basis function.

such shape functions,

$$u_h(x) = \sum_{j=1}^{3} u_j e_j(x) \quad \text{or, more precisely,} \quad u_h|_K(x) = \sum_{j=1}^{3} u_j e_j|_K(x) \,.$$

Note that d.o.f. u_j can be identified as the value of u_h at vertex v_j and interpreted as a linear (Dirac) functional returning for a function u_h its value at vertex v_j,

$$\langle \psi_j, u_h \rangle = \psi_j(u_h) := u_h(v_j) \,.$$

If we consider a globally continuous function u, and set $u_j = \psi_j(u) = u(v_j)$ above, we obtain the *piecewise* linear interpolant of u,

$$\Pi_h u = \sum_j \psi_j(u) e_j = \sum_j u(v_j) e_j \,.$$

Operator Π_h, prescribing for each continuous function u its interpolant $\Pi_h u$, is identified as the *interpolation operator* corresponding to the Courant triangle,

$$C(\overline{\Omega}) \ni u \to \Pi_h u \in X_h \,,$$

where

$$X_h = \text{span}\{e_j\} \subset H^1(\Omega)$$

is the *FE approximation space*. We introduce the corresponding concepts for element K: *space of shape functions* $X_h(K)$, element d.o.f. $\psi_{j,K}$, and element interpolation operator $\Pi_{h,K}$,

$$X_h(K) := \mathcal{P}^1(K) = \text{span}\left\{\phi_{j,K}\right\},$$

$$\psi_{j,K}(u) = u(v_{j,K}), \quad j = 1, 2, 3 \,,$$

$$\Pi_{h,K} u := \sum_{j=1}^{3} \psi_{j,K}(u) \phi_{j,K} = \sum_{j=1}^{3} u(v_{j,K}) \phi_{j,K}$$

with $u \in C(\overline{K})$, element vertices $v_{j,K}$, and element shape functions $\phi_{j,K}$.

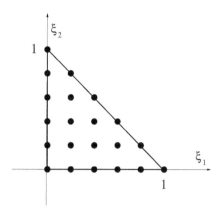

Figure 3.2: Lagrange triangle of order $p = 5$.

Lagrange triangle of order p. The ideas behind the Courant triangle can be easily generalized to the Lagrange triangle of arbitrary order $p \geq 1$. We begin by introducing a set of uniformly distributed *Lagrange nodes*. Fig. 3.2 presents the Lagrange nodes for the case of $p = 5$ and a unit (right) triangle K. First of all, notice that the number of Lagrange nodes coincides with the dimension of polynomial space $\mathcal{P}^5(K)$ (just count the number of monomials in the Pascal triangle). With each Lagrange node a_j we associate the corresponding d.o.f. returning the function value at the node,

$$\psi_j(u) = u(a_j) \,.$$

Consequently, the jth Lagrange shape function will be a polynomial of order 5 taking on value one at a_j and vanishing at the remaining nodes. Take time to write explicit formulas for the Lagrange shape functions corresponding to selected nodes. We can classify the shape functions into three groups:

- vertex shape functions corresponding to nodes at the three vertices,
- *edge bubbles* corresponding to nodes in (the interior of) an edge,
- *element bubbles* corresponding to nodes in (the interior of) the element.

Note that there are $p - 1$ edge bubbles for each edge, and $(p - 2)(p - 1)/2$ element bubbles. The element shape functions are now glued into basis functions. Element vertex shape functions contribute to *vertex basis functions* spanning over all elements adjacent to a vertex. Edge bubbles contribute to *edge basis functions* with supports spanning at most two elements (for edges on the boundary, the support will consist of a single element only). And finally, element bubbles are extended by zero to yield global element bubble basis functions. Note that all basis functions are globally continuous (explain why). The element interpolation operator is given by

$$C(\overline{K}) \ni u \longrightarrow \Pi_{h,K} u = \sum_{j=1}^{n} \psi_j(u)\phi_j = \sum_{j=1}^{n} u(a_j)\phi_j \,,$$

where $n = (p + 1)(p + 2)/2$.

3.1.2 ▪ Ciarlet's Definition of a Finite Element

The ideas discussed so far were generalized by Ciarlet [18] to an abstract concept of a finite element. In order to define a finite element, we must introduce

- geometry of the element, usually a polygon or polyhedral K;

- a space of FE shape functions (usually polynomials) $X(K)$ contained in the appropriate energy space, dim $X(K) = n$;

- a set of linear and continuous functionals ψ_j, called *degrees of freedom* (d.o.f.), defined on a subset $\mathcal{X}(K)$ (of sufficiently regular functions) of the energy space containing the FE space $X(K)$,

$$\psi_j \,:\, \mathcal{X}(K) \to \mathbb{R}(\mathbb{C}), \quad j = 1, \ldots, n, \tag{3.1.1}$$

  such that restrictions of ψ_j to $X(K)$ are linearly independent, i.e., they form a basis in the algebraic dual of $X(K)$.

The linear independence condition is known as the *unisolvence condition*. The corresponding dual basis in $X(K)$,

$$\phi_i \in X(K), \quad \langle \psi_j, \phi_i \rangle = \delta_{ij}, \quad i, j = 1, \ldots, n, \tag{3.1.2}$$

is identified as *FE shape functions*. The following is a useful characterization of the unisolvence condition.

Lemma 3.1. *The following conditions are equivalent to each other:*

(i) *Restrictions $\psi_j|_{X(K)}$, $j = 1, \ldots, n$, are linearly independent.*

(ii) *Vanishing of all d.o.f. implies vanishing of the shape function,*

$$\psi_j(\phi) = 0, \quad j = 1, \ldots, n, \quad \Rightarrow \quad \phi = 0, \qquad \phi \in X(K). \tag{3.1.3}$$

Proof. (i) $\Rightarrow$ (ii) follows from the fact that ψ_j, $j = 1, \ldots, n$, span the algebraic dual.
(ii) $\Rightarrow$ (i). Condition (3.1.3) implies that ψ_j, $j = 1, \ldots, n$, span the algebraic dual. As their number matches the dimension of the space, they must be linearly independent. $\square$

The definition should be treated rather informally. It tells only a part of the story. In particular, in the case of H^1-conforming elements, implicit in the construction is an assumption that by equating certain d.o.f. for neighboring elements, we guarantee that the union of the FE shape functions lives in the global energy space. This is best explained starting with examples.

Lagrange master finite elements.

- *Element:* simplicial elements: master interval I, triangle and tetrahedron T; tensor product elements: master quad I^2, master hexa (cube) I^3; master prism: $T \times I$.

- The corresponding *FE spaces of shape functions* are

$$\mathcal{P}^p(K) \quad \text{(simplices)},$$

$$\mathcal{Q}^{p,q} := \mathcal{P}^p \otimes \mathcal{P}^q \quad \text{(quad)},$$

$$\mathcal{Q}^{p,q,r} := \mathcal{P}^p \otimes \mathcal{P}^q \otimes \mathcal{P}^r \quad \text{(cube)},$$

$$\mathcal{P}^p(T) \otimes \mathcal{P}^q(I) \quad \text{(prism)},$$

  where p, q, r denote the polynomial order in one, two, or three space dimensions.

- *Degrees of freedom:* values of shape functions at the Lagrangian nodes:

$$\psi_j \,:\, \mathcal{X}(K) = C(\overline{K}) \ni \phi \to \phi(a_j) \in \mathbb{R}.$$

Lagrangian nodes are uniformly distributed over the master element; their number matches the dimension of the corresponding space of element shape functions. I am frequently drawing them to compute the dimension of the space.

3.1.3 ▪ Parametric H^1-Conforming Lagrange Element

The concept of Lagrange element can be extended to elements of arbitrary shape, possibly curvilinear. Given a master element $\hat{K}$ and an element map x_K from $\hat{K}$ *onto* a *physical* element $K \subset \mathbb{R}^N$,

$$x_K : \hat{K} \to K, \quad x = x(\xi),$$

we introduce the triple:

- element K,

- space of element shape functions,

$$X(K) := \{\hat{u} \circ x_K^{-1} : \hat{u} \in X(\hat{K})\},$$

- element d.o.f.,

$$\psi_j : X(K) \ni u \to u(a_j) \in \mathbb{R},$$

  where a_j is the image of Lagrangian node $\hat{a}_j$ in the master element.

Note the commuting property:

$$\langle \psi_j, u \rangle = \langle \hat{\psi}_j, \hat{u} \rangle.$$

For general parametric elements, the commuting property may be enforced by definition, i.e., it defines the d.o.f. on the physical element. Note that the parametric element shape functions, in general, are *not* polynomials. Only in the case of an affine element map is its inverse also an affine map and, therefore, compositions of the affine map with polynomials remain polynomials. We speak then about an *affine finite element*.

Element interpolation operator. The interpolation operator is constructed according to Ciarlet's definition,

$$\mathcal{X}(K) \ni u \to \Pi_K u := \sum_j^n \langle \psi_j, u \rangle \phi_j \in X(K).$$

The commuting property for the d.o.f. implies the corresponding commuting property for interpolation on master and physical elements,

$$(\Pi_K u) \circ x_K = \hat{\Pi}_K (u \circ x_K),$$

or, in a more concise form ("breaking the hat" property),

$$\widehat{(\Pi_K u)} = \hat{\Pi}_{\hat{K}} \hat{u}.$$

We can illustrate the property in terms of the commuting diagram

$$
\begin{array}{ccc}
u & \xrightarrow{\Pi_K} & \Pi_K u \\[1mm]
\downarrow x_K^{-1} & & \downarrow x_K^{-1} \\[1mm]
\hat{u} & \xrightarrow{\hat{\Pi}_{\hat{K}}} & \hat{\Pi}_{\hat{K}} \hat{u} = \widehat{\Pi_K u}.
\end{array}
\qquad (3.1.4)
$$

Global finite element space and global conformity.

$$X_h := \{u \in H^1(\Omega) \ : \ u|_K \in X(K) \quad \forall K \in \mathcal{T}_h\}\,.$$

Conformity or, equivalently, global continuity of functions from the FE space is implied by two assumptions:

- conformity of master finite elements implying immediately conformity of affine elements,

- *global continuity* of element maps.

Let us start with the 2D case of two affine elements K_1, K_2 sharing a common edge e. We assume that the shape functions for both elements are polynomials of the same order p along the common edge. We can match, of course, two triangles of the same order, but we can also match a quad space $\mathcal{Q}^{p,q}$ with a triangle space $\mathcal{P}^p$ provided the orders along the common edge are equal, say, p_e. This implies that both elements share two vertex nodes and $p_e - 1$ Lagrangian nodes in the interior of the common edge. Equating values at the common $p_e + 1$ nodes then implies the global continuity along the edge. Once the conformity of affine elements has been confirmed, and the element maps coincide with each other on the common edge, the conformity of parametric elements follows.

I am going to repeat now the same arguments in a more formal way by introducing the concept of a *finite subelement* in the spirit of Ciarlet's definition. Let S denote a face or edge (or vertex) of a finite element K, with the corresponding finite element subspace $X_h(S)$ and d.o.f. $\psi_{j,S}, j = 1, \ldots, \dim X_h(S)$. We say that triple $(S, X_h(S), \psi_{j,S})$ is a *subelement of element* $(K, X_h(K), \psi_{j,K})$ if the following conditions are satisfied:

- S is a face or edge of K.

- Space $X_h(S)$ coincides with the space of restrictions:

$$X_h(S) := \{u_h|_S \ : \ u_h \in X_h(K)\}\,.$$

- For every d.o.f. $\psi_{j,S}, j = 1, \ldots, \dim X_h(S)$, there exists a unique element d.o.f. $\psi_{j,K}$ such that, for every $u_h \in X_h(S)$,

$$\psi_{j,S}(u_h) = \psi_{j,K}(U_h)\,,$$

  where $U_h \in X_h(K)$ is any extension of u_h. In particular, the value of $\psi_{j,K}(U_h)$ is independent of the extension.

Note that the subelement is unique up to a possible renumeration of its d.o.f. A finite element mesh is now globally conforming if, for any two adjacent elements, sharing a vertex, edge, or face S, there exists a common restriction $(S, X_h(S), \psi_{j,S})$ of the two elements. For a parametric subelement, this implies that there exists a subelement map x_S mapping a reference element $\hat{S}$ onto S.

Remark 3.2. It is possible to match two elements along a common edge or face even if they do not share a common subelement by means of *constrained approximation*. *Constrained element* d.o.f. must be expressed as linear combinations of corresponding *parent (unconstrained) element* d.o.f. Such matching requires a nonstandard assembly procedure; see [25, 35] for details.

Enforcement of global continuity leads to the identification of d.o.f. for element vertices, edges, and faces, their equality for neighboring elements, and, eventually, the notion of d.o.f. as well-defined functionals on the global FE space X_h,

$$\psi_j : \mathcal{X}(\Omega) \supset X_h \to \mathbb{R}\,.$$

The corresponding dual basis is identified as the (Galerkin) *basis functions* e_i. Degrees-of-freedom and the basis functions are naturally classified into vertex, edge, face, and element interior d.o.f. and basis functions. The basis functions are unions of the corresponding element shape functions. Finally, we have the global interpolation operator:

$$\Pi_h u = \sum_j \langle \psi_j, u \rangle e_j \, .$$

Both symbols for the global and element interpolation operator Π_h and global and local d.o.f. ψ_j are typically overloaded skipping symbols K and $\hat{K}$ for the physical or master element, respectively.

Isoparametric, subparametric, and superparametric elements. If the element map x_K lives in the master element space of shape functions, i.e., it can be represented in the form

$$x_K(\xi) = \sum_j x_{K,j} \hat{\phi}_j(\xi) \, ,$$

we speak about an *isoparametric finite element*. Vector-valued coefficients $x_{K,j}$ are identified as *geometry d.o.f.* and have to be defined during mesh generation. For Lagrange elements the geometry d.o.f. $x_{K,j}$ are simply coordinates of the corresponding Lagrange nodes a_j. The idea of an isoparametric FE element is usually credited to Irons, Ergatoudis, and Zienkiewicz [52, 42]. Isoparametric elements have a simple but remarkable property: the element space of shape functions $X_h(K)$ always contains linear polynomials,

$$\mathcal{P}^1(K) \subset X_h(K) \, ;$$

see Exercise 3.1.6. For linear elasticity, this translates into the observation that the global FE space includes *linearized rigid body motions*. Given the fact that, in general, shape functions of an isoparametric element are not polynomials, this is a remarkable property. Among other things, we can use it to verify a finite element code. No matter how curvilinear and what order the mesh is, one must be able to reproduce global linear functions with machine precision. This property is not limited to polynomials—it remains true as long as the element maps live in the master element space of (possibly nonpolynomial) shape functions. For example, see the concept of *isogeometric discretizations* using splines and NURBS [22].

If the element map comes from a proper subspace of the element space of shape functions $X_h(K)$, we talk about a *subparametric element*. Affine elements are an example of subparametric elements. Finally, if the element map comes from a proper superspace of the element space of shape functions $X_h(K)$, we talk about a *superparametric element*. The superparametric elements are used when one is concerned with the geometry approximation error like in the *boundary element* method or interface problems. In particular, the *exact geometry element* [25, 35] can be formally classified as a superparametric element as well.

3.1.4 ▪ Hierarchical Shape Functions

In the p-version of the FE method, the mesh is fixed and we converge to the exact solution by raising the polynomial order p of approximation, hence the name. The order can be raised *uniformly* or *adaptively*, i.e., only in some elements. We arrive at the need of meshes combining elements of *varying order*. The use of such elements takes place also in the h-version of the FE method. Frequently, we need to employ higher order elements ($p = 4, 5$) locally[6] with most of

---

[6]For example, to avoid the so-called shear or membrane locking phenomenon occurring in the discretization of thin-walled structures.

the domain discretized with lower order elements, say, $p = 2$, hence the need for building a code that supports *variable order* elements.

Varying polynomial order is practically infeasible with Lagrange elements, but it is very natural and straightforward with *hierarchical shape functions* that have been used in the p-method of Barna Szabo from the very beginning; see the preface in [25] for a detailed historical account.

The rise of the p-method and hierarchical shape functions revealed limitations of Ciarlet's formalism for constructing finite elements. Hierarchical shape functions (Szabo called them *modes*) reflected the geometry of the finite element mesh and were constructed without defining d.o.f. first. They are classified into vertex, edge, face, and element shape functions. Support of a vertex basis function spans over all elements sharing the vertex, support of an edge basis function consists of all elements sharing the edge, support of a face basis function spans over (at most two) elements sharing the face, and, last, support of an element basis function includes the element only. As for Lagrange elements, the basis functions are unions of the respective contributing element shape functions, possibly premultiplied with a sign factor accounting for orientation.

Once the shape functions have been introduced, we may try to identify the corresponding d.o.f. and then proceed with the construction of the interpolation operator. This is not so straightforward as the shape functions imply the uniqueness of the d.o.f. (the dual basis) only on the FE space of element shape functions $X(K)$ but not the bigger and rather ambiguous subspace $\mathcal{X}(K)$ of the energy space. Recall that, in the Ciarlet definition, the choice of subspace $\mathcal{X}(K)$ is simply driven by necessary regularity assumptions to make the d.o.f. well-defined. Early attempts to identify d.o.f. corresponding to hierarchical shape functions led to wrong choices of subspace $\mathcal{X}(K)$ and, most importantly, suboptimal interpolation operators.

An alternative came with the construction of *projection-based (PB) interpolation* operators, [65, 29, 24, 14, 30, 26]. Here, we construct the interpolation operators *without* using any d.o.f.—in fact, we do not need to define the d.o.f. at all. Out of academic curiosity, we may try to identify d.o.f. that would result in the PB interpolation using Ciarlet's definition.

Conformity with hierarchical shape functions. Enforcing conformity with hierarchical shape functions is relatively straightforward. We begin by introducing vertex basis functions. A shape function for the 0-dimensional vertex is just a scalar equal one. Consider a particular vertex in the mesh. For each edge adjacent to the vertex, we extend the scalar to a linear function vanishing at the other end of the edge—the linear vertex edge shape function.

We proceed with edge basis functions. For each edge, we introduce the *edge system of coordinates* ξ_e with 1D shape functions defined on the edge, including the two linear vertex shape functions just introduced, and *edge bubbles*. Given the edge shape functions, for each adjacent face, we extend them into adjacent face shape functions. These extensions use minimum order polynomials (on the master element) and have to vanish on all other edges. The edge coordinate ξ_e may or may not coincide with the corresponding *local* edge coordinate implied by the face system of coordinates in which the extension is calculated. If the face shape function matches the extension of the edge shape function, we talk about the *orientation-embedded* shape functions [47]. The original element shape functions of Barna Szabo matched the edge shape functions up to a multiplicative factor ± 1 that had to be taken into account during the assembly procedure.

We proceed then in the same way with face basis functions. Each face is equipped with global face coordinates ξ_f and corresponding 2D face functions, including the extensions of vertex shape functions and edge bubbles, as well as newly introduced *face bubbles*. The face functions must be extended into the neighboring elements. Orientation embedding for faces is much more important than for edges. Without it, generation of hierarchical bubble shape functions for triangular faces and arbitrary element systems of coordinates is impossible; see the discussion in [35, p. 50].

Finally, we construct *element bubbles*, i.e., basis functions whose support spans a single element only.

The logic of constructing global basis functions extends to the construction of element maps for parametric elements. We begin by introducing vertices. Then, for each edge, we construct a parametrization on a unit master interval $\hat{I} = (0, 1)$ that matches the endpoint vertex coordinates. In the next step, for each triangular or quadrilateral face, we construct a parametrization mapping the corresponding master triangle or square onto the physical space. The parametrization must be *compatible* with already existing parametrizations for the face edges. A number of techniques, including *transfinite parametrizations* or *implicit parametrizations*, can be used; see [25, 35] for details. In the last step, we extend the face parametrizations to element parametrizations using the same techniques. The "bottom-up" approach enforces the global continuity of element maps, which, in turn, guarantees global conformity of parametric elements.

Exercises

3.1.1. Lagrange square element.

(i) Draw the master quad of order $(3, 4)$ and the corresponding Lagrange nodes. Use elementary means to construct shape functions for a sample vertex, edge, and interior node. Check that they are in the space of element shape functions.

(ii) Use the Lagrange shape functions to prove the *unisolvence condition*. Note that this approach is mathematically awkward as the shape functions are supposed to be defined *after* the unisolvency is established. Can you think of alternate ways to prove the unisolvency *without* using the shape functions?

(iii) Think of possible ways to modify the location of the Lagrangian nodes to keep the unisolvency condition intact.

(iv) Consider two master elements sharing an edge and assume the order of the elements in such a way that the restrictions of element shape functions to the common edge live in the same polynomial space. Explain why matching the d.o.f. (pointwise values) at the common edge Lagrangian nodes implies global continuity of functions obtained by "gluing" shape functions defined on the two elements (the mathematical term is *unions* of shape functions).

(v) Going back to the question asked in step (iii), is the location of Lagrangian nodes and their number at vertices, edges, and interior essential for enforcing the global continuity? Discuss possible modifications to the Lagrangian nodes that would preserve global continuity.

(5 points)

3.1.2. Lagrange triangular element. Repeat the steps of Exercise 3.1.1 for the master triangle of order 5 shown in Fig. 3.2. (3 points)

3.1.3. Lagrange 3D element. Pick your favorite 3D element and repeat for it the steps of Exercise 3.1.1. (3 points)

3.1.4. Parametric Lagrange element.

(i) Prove the unisolvency for an arbitrary parametric Lagrange element. Discuss why it is necessary for the element map to remain bijective in the element closure $\bar{\hat{K}}$ (this eliminates the possibility of singular maps like Duffy's map).

(ii) Assume you have two physical 2D Lagrange elements K_1, K_2 sharing an edge e. Let x_{K_i} be the corresponding element maps defined on master elements $\hat{K}_i$, $i = 1, 2$. Discuss sufficient conditions on master element space $X(\hat{K}_i)$ and the element maps that would guarantee the global continuity of unions of FE shape functions.

(5 points)

3.1.5. **Alternate d.o.f.** Consider your favorite 3D Lagrangian element of arbitrary order and replace the Lagrangian d.o.f. with a new set of d.o.f. defined by using edge, face, and element moments:

$$\text{vertex d.o.f. : } u \to (v) \qquad\qquad \forall \text{ vertex } v,$$

$$\text{edge d.o.f. : } u \to \int_e u f_i^e, \quad i = 1, \dots, ?, \quad \forall \text{ edge } e,$$

$$\text{face d.o.f. : } u \to \int_f u f_i^f, \quad i = 1, \dots, ?, \quad \forall \text{ face } f,$$

$$\text{interior d.o.f. : } u \to \int_K u f_i^K, i = 1, \dots, ?.$$

Discuss the number of edge, face, and interior moments necessary for enforcing the global continuity and satisfying the unisolvence condition. Provide a concrete example of weights f_i^e, f_i^f, f_i^K with which the element satisfies the unisolvency condition. (5 points)

3.1.6. Prove that, for any isoparametric finite element, the element space of shape functions $X_h(K)$ always contains linear polynomials,

$$\mathcal{P}^1(K) \subset X_h(K).$$

(5 points)

3.2 ▪ Exact Sequence Elements

In this section, we extend the H^1-conforming elements studied in Section 3.1 to a family of elements forming the *exact grad-curl-div sequence*. Recall from functional analysis that a sequence of vector spaces X_i, $i = 0, \dots, n$, and corresponding linear operators $A_i : X_{i-1} \to X_i$, $i = 1, \dots, n$, is said to be an *exact sequence* if the range of each operator coincides with the null space of the next operator, i.e.,

$$\mathcal{R}(A_i) = \mathcal{N}(A_{i+1}), \quad i = 1, \dots, n-1.$$

For an open set $\Omega \subset \mathbb{R}^3$ homeomorphic with an open ball, operators grad-curl-div and the energy spaces introduced in Chapter 1 form the exact sequence

$$\mathbb{R} \xrightarrow{\text{id}} H^1(\Omega) \xrightarrow{\nabla} H(\text{curl}, \Omega) \xrightarrow{\nabla\times} H(\text{div}, \Omega) \xrightarrow{\nabla\cdot} L^2(\Omega) \xrightarrow{0} \{0\}.$$

Above, the symbol $\mathbb{R}$ stands for *constant functions* and "id" is the identity operator. The first segment of the exact sequence communicates thus only that the null space of grad operator is formed by constant functions. Similarly, the last trivial space and trivial operator communicate only that the div operator is surjective. Keeping these two facts in mind, we shorten the exact sequence to

$$H^1(\Omega) \xrightarrow{\nabla} H(\text{curl}, \Omega) \xrightarrow{\nabla\times} H(\text{div}, \Omega) \xrightarrow{\nabla\cdot} L^2(\Omega).$$

The sequence communicates now two additional important properties of grad, curl, and div operators,

$$E \in H(\mathrm{curl}, \Omega), \ \boldsymbol{\nabla} \times E = 0 \quad \Leftrightarrow \quad \text{there exists a function (scalar potential)}$$

$$u \in H^1(\Omega) \ : \ \boldsymbol{\nabla} u = E \,,$$

$$v \in H(\mathrm{div}, \Omega), \ \boldsymbol{\nabla} \cdot v = 0 \quad \Leftrightarrow \quad \text{there exists a function (vector potential)}$$

$$E \in H(\mathrm{curl}, \Omega) \ : \ \boldsymbol{\nabla} \times E = v \,.$$

Note that the scalar potential is unique up to an additive constant but the vector potential is unique only up to a gradient; recall the role of various *gauge conditions* in electromagnetics to make it unique.

For a general domain Ω, we can claim only that

$$\boldsymbol{\nabla} \times (\boldsymbol{\nabla} u) = 0 \quad \Rightarrow \quad \mathcal{R}(\mathrm{grad}) \subset \mathcal{N}(\mathrm{curl}) \,,$$

$$\boldsymbol{\nabla} \cdot (\boldsymbol{\nabla} \times E) = 0 \quad \Rightarrow \quad \mathcal{R}(\mathrm{curl}) \subset \mathcal{N}(\mathrm{div}) \,.$$

We talk then only about the *differential complex*.

3.2.1 ▪ Polynomial Exact Sequences

As we have learned in the previous section, finite elements can be defined in different ways following the logic of Ciarlet (d.o.f. first) or Szabo (shape functions first), but in either case we have to specify first the discrete finite element spaces: local FE space of element shape functions $X_h(K)$, and global FE space X_h. In this section, we will seek discrete polynomial (locally) and piecewise polynomial (globally) subspaces of the energy spaces that reproduce the algebraic structure of the exact grad-curl-div sequence on the discrete level.

3D exact sequence.

$$H^1 \xrightarrow{\boldsymbol{\nabla}} H(\mathrm{curl}) \xrightarrow{\boldsymbol{\nabla} \times} H(\mathrm{div}) \xrightarrow{\boldsymbol{\nabla} \cdot} L^2$$

$$\cup \qquad\quad \cup \qquad\qquad \cup \qquad\quad \cup \tag{3.2.5}$$

$$W^p \xrightarrow{\boldsymbol{\nabla}} \quad Q^p \xrightarrow{\boldsymbol{\nabla} \times} \quad V^p \xrightarrow{\boldsymbol{\nabla} \cdot} Y^p \,.$$

Symbols W^p, Q^p, V^p, Y^p, introduced by Doug Arnold, will stand (loosely ...) for both element and global FE spaces. Index p is supposed to indicate different polynomial degrees and should not be interpreted literally. For instance, for the so-called first Nédélec sequence (of discrete spaces), W^p will contain complete polynomials of order p, but the remaining spaces Q^p, V^p, Y^p will contain complete polynomials of order $p - 1$ only.

The 3D sequence gives rise to two 2D sequences and a 1D sequence. We start with two possible 2D scenarios for the computation of the curl.

Case: $E = (E_1, E_2, 0)$, $E_1 = E_1(x, y)$, $E_2 = E_2(x, y)$,

$$\boldsymbol{\nabla} \times E = (0, 0, E_{2,1} - E_{1,2})$$

leads to the definition

$$E = (E_1, E_2), \quad \mathrm{curl}\, E := E_{2,1} - E_{1,2} \,.$$

Case: $E = (0, 0, E_3)$, $E_3 = E_3(x, y)$,

$$\boldsymbol{\nabla} \times E = (E_{3,2}, -E_{3,1}, 0)$$

leads to the definition

$$u = u(x, y), \quad \boldsymbol{\nabla} \times u = \left(\frac{\partial u}{\partial y}, -\frac{\partial u}{\partial x}\right).$$

The two 2D exact sequences with their discrete counterparts look as follows.

2D exact sequence:

$$\begin{array}{ccccc}
H^1 & \xrightarrow{\boldsymbol{\nabla}} & H(\mathrm{curl}) & \xrightarrow{\mathrm{curl}} & L^2 \\
\cup & & \cup & & \cup \\
W^p & \xrightarrow{\boldsymbol{\nabla}} & Q^p & \xrightarrow{\mathrm{curl}} & Y^p.
\end{array} \quad (3.2.6)$$

"Rotated" 2D exact sequence:

$$\begin{array}{ccccc}
H^1 & \xrightarrow{\boldsymbol{\nabla}\times} & H(\mathrm{div}) & \xrightarrow{\mathrm{div}} & L^2 \\
\cup & & \cup & & \cup \\
W^p & \xrightarrow{\boldsymbol{\nabla}\times} & V^p & \xrightarrow{\mathrm{div}} & Y^p.
\end{array} \quad (3.2.7)$$

We finish with the simplest 1D case.

1D exact sequence:

$$\begin{array}{ccc}
H^1 & \xrightarrow{\partial} & L^2 \\
\cup & & \cup \\
W^p & \xrightarrow{\partial} & Y^p,
\end{array} \quad (3.2.8)$$

where symbol ∂ stands for the derivative. The element spaces in the 1D case are unique, $W^p = \mathcal{P}^p$, and $Y^p = \mathcal{P}^{p-1}$. We shall present several possible constructions for 2D and 3D discrete sequences.

3.2.2 ▪ Lowest Order Elements and Commuting Interpolation Operators

Along with the spaces, we will seek the construction of corresponding *commuting interpolation operators* that can be constructed through d.o.f. or directly, through local projections.

Lowest order tetrahedral element of the first type.   Let K be an arbitrary tetrahedron. The FE spaces are defined as follows:

$$\begin{aligned}
W &= W^1 = \mathcal{P}^1(K), \\
Q &= Q^1 = \{E \in \mathcal{P}^1(K)^3 \,:\, E_t|_e \in \mathcal{P}^0(e) \text{ for each edge } e\}, \\
V &= V^1 = \{v \in \mathcal{P}^1(K)^3 \,:\, v_n|_f \in \mathcal{P}^0(f) \text{ for each face } f\}, \\
Y &= Y^1 = \mathcal{P}^0(K),
\end{aligned} \quad (3.2.9)$$

where $E_t = E \cdot \tau_e$ is the tangential component of vector E, and $v_n = v \cdot n_f$ is the normal component of v with τ_e denoting a unit tangent vector for edge e, and n_f a unit normal vector for face f. Note that the definition is independent of the choice of the edge and face unit vectors, and

$$\dim W = \text{ number of vertices } = 4\,,$$
$$\dim Q = \text{ number of edges } = 6\,,$$
$$\dim V = \text{ number of faces } = 4\,,$$
$$\dim Y = \text{ number of elements } = 1\,.$$

The element d.o.f. can be defined as follows:

$$H^1(K) \supset ? \ni u \to u(v) \in \mathbb{R} \qquad \text{for each vertex } v\,,$$

$$H(\text{curl}, K) \supset ? \ni E \to \int_e E_t \in \mathbb{R} \qquad \text{for each edge } e\,,$$

$$H(\text{div}, K) \supset ? \ni v \to \int_f v_n \in \mathbb{R} \qquad \text{for each face } f\,, \qquad (3.2.10)$$

$$L^2(K) \ni q \to \int_K q \in \mathbb{R} \qquad \text{for element } K\,.$$

The tangential and normal components are defined using specific tangential edge and face normal unit vectors. The question marks stand for subspaces of energy spaces, consisting of sufficiently regular functions for which the d.o.f. are well-defined. They are usually characterized in terms of Sobolev spaces H^s with real exponent, $s \in \mathbb{R}$. We shall specify them later after we review some fundamental facts about Sobolev spaces.

The interpolation operators $\Pi^{\text{grad}}, \Pi^{\text{curl}}, \Pi^{\text{div}}$ corresponding to the d.o.f. can be equivalently specified as unique operators satisfying the conditions

$$\Pi^{\text{grad}} u - u = 0 \text{ at each vertex } v\,,$$

$$\int_e (\Pi^{\text{curl}} E - E)_t = 0 \text{ for each edge } e\,, \qquad (3.2.11)$$

$$\int_f (\Pi^{\text{div}} v - v) \cdot n_f = 0 \text{ for each face } f\,.$$

Note the independence of the interpolation operators from the selected tangential and normal unit vectors. The interpolation operator for the L^2-spaces is simply the L^2-projection, and the property

$$\int_K (Pq - q) = 0$$

is an equivalent definition of L^2-projection onto constants.

Finally, note that the discussed d.o.f. guarantee not only the unisolvence conditions but the conformity of the global discretization as well. If you miss this fact, you are in big trouble.

Whitney shape functions. Let a_0, a_1, a_2, a_3 denote the vertices of a tetrahedron. Vectors $a_i - a_0$, $i = 1, 2, 3$, are linearly independent and, therefore, for each point $x \in \mathbb{R}^3$, there exist unique numbers (components) λ_i, $i = 1, 2, 3$, such that

$$x - a_0 = \sum_{i=1}^{3} \lambda_i (a_i - a_0)$$

or, equivalently,

$$x = \underbrace{(1 - \lambda_1 - \lambda_2 - \lambda_3)}_{=:\lambda_0} a_0 + \lambda_1 a_1 + \lambda_2 a_2 + \lambda_3 a_3\,.$$

Numbers $\lambda_0, \ldots, \lambda_3$ are identified as the *affine (barycentric) coordinates* of point x with respect to the (vertices of) tetrahedron K. One can show that λ_i are linear functions of x and that they are invariant under affine isomorphisms; compare Exercise 3.2.5.

The following *Whitney shape functions* form bases for the lowest order tetrahedron of the first type:

$$\lambda_i, \qquad\qquad\qquad\qquad\qquad\qquad\qquad i = 0, 1, 2, 3\,,$$

$$\lambda_i \nabla \lambda_j - \lambda_j \nabla \lambda_i, \qquad (i, j) = (0, 1), (1, 2), (0, 2), (0, 3), (1, 3), (2, 3)\,,$$

$$\lambda_i (\nabla \lambda_j \times \nabla \lambda_k) + \lambda_k (\nabla \lambda_i \times \nabla \lambda_j) + \lambda_j (\nabla \lambda_k \times \nabla \lambda_i)\,,$$

$$(i, j, k) = (0, 1, 2), (0, 1, 3), (1, 2, 3), (0, 2, 3)\,,$$

$$\frac{1}{|K|} \qquad\qquad\qquad\qquad\qquad\qquad \text{(constant function)}\,,$$

where $|K|$ is the volume of tetrahedron K. The shape functions correspond to the following d.o.f. (see Exercise 3.2.6):

- H^1 element:

$$\phi \to \phi(a_i), \quad i = 0, 1, 2, 3\,.$$

- $H(\text{curl})$ element:

$$E \to \frac{1}{|e_{ij}|} \int_{e_{ij}} E \cdot (a_j - a_i), \quad (i, j) = (0, 1), (1, 2), (0, 2), (0, 3), (1, 3), (2, 3)\,,$$

where e_{ij} denotes the edge from vertex a_i to vertex a_j, and $|e_{ij}| = |a_j - a_i|$ stands for its length. Note that

$$\tau_e = \frac{a_j - a_i}{|e_{ij}|}$$

is the edge unit vector.

- $H(\text{div})$ element:

$$v \to \frac{1}{|f_{ijk}|} \int_{f_{ijk}} v \cdot [(a_j - a_i) \times (a_k - a_i)], \quad (i, j, k) = (0, 1, 2), (0, 1, 3), (1, 2, 3), (0, 2, 3)\,,$$

where f_{ijk} denotes the face spanned by vertices a_i, a_j, a_k, and $|f_{ijk}| = |(a_j - a_i) \times (a_k - a_i)|$ is the area of the face. Note that face normal unit vector n_f is given by

$$n_f = \frac{(a_j - a_i) \times (a_k - a_i)}{|(a_j - a_i) \times (a_k - a_i)|}\,.$$

- L^2 element:

$$q \to \int_K q\,.$$

Note that the d.o.f. coincide with those defined earlier in (3.2.10).

It is illuminating to express gradients $\nabla \lambda_j$ and products of gradients $\nabla \lambda_j \times \nabla \lambda_k$ in the formulas for Whitney shape functions in terms of basis and co-basis vectors corresponding to affine coordinates λ_i, $i = 1, 2, 3$,

$$x = a_0 + \sum_{i=1}^{3} \lambda_i \underbrace{(a_i - a_0)}_{=: g_i}\,.$$

Let g^j be the co-basis of g_i,

$$g^1 = \frac{g_2 \times g_3}{[g_1, g_2, g_3]}, \quad g^2 = \frac{g_3 \times g_1}{[g_1, g_2, g_3]}, \quad g^3 = \frac{g_1 \times g_2}{[g_1, g_2, g_3]}\,,$$

where $[g_1, g_2, g_3] = g_1 \cdot (g_2 \times g_3) = |K|$. Recalling the formula for gradient ∇u of a function $u = u(x)$ in a curvilinear system of coordinates [35, Appendix 1],

$$\nabla u = \sum_{i=1}^{3} \frac{\partial u}{\partial \lambda_i} g^i,$$

we realize that

$$\nabla \lambda_i = g^i, \quad i = 1, 2, 3.$$

Similarly,

$$\nabla \lambda_i \times \nabla \lambda_j = g^i \times g^j = [g^1, g^2, g^3] g_k = [g_1, g_2, g_3]^{-1} g_k$$

for any cyclic permutation $[i, j, k]$ of $1, 2, 3$.

Invariance of affine coordinates with respect to affine isomorphisms implies that the Whitney formulas remain valid *for any tetrahedron K.*

De Rham diagram. Commutativity of interpolation operators. The following de Rham diagram communicates commuting properties of the interpolation operators.

$$
\begin{array}{ccccccc}
H^1 & \xrightarrow{\nabla} & H(\mathrm{curl}) & \xrightarrow{\nabla\times} & H(\mathrm{div}) & \xrightarrow{\nabla\cdot} & L^2 \\
\downarrow \Pi^{\mathrm{grad}} & & \downarrow \Pi^{\mathrm{curl}} & & \downarrow \Pi^{\mathrm{div}} & & \downarrow P \\
W^p & \xrightarrow{\nabla} & Q^p & \xrightarrow{\nabla\times} & V^p & \xrightarrow{\nabla\cdot} & Y^p,
\end{array}
\qquad (3.2.12)
$$

where $\Pi^{\mathrm{grad}}, \Pi^{\mathrm{curl}}, \Pi^{\mathrm{div}}$ are the interpolation operators and P denote the L^2-projection.

Theorem 3.3. *The FE spaces corresponding to the lowest order tetrahedron of the first type and the corresponding interpolation operators satisfy the de Rham diagram.*

Proof. We start with the commutativity of Π^{grad} and Π^{curl},

$$\nabla(\Pi^{\mathrm{grad}} u) \overset{?}{=} \Pi^{\mathrm{curl}}(\nabla u).$$

As both sides live in space Q^1, by the shape functions reproducibility property, the statement is equivalent to

$$\Pi^{\mathrm{curl}}\left(\nabla(\Pi^{\mathrm{grad}} u)\right) = \Pi^{\mathrm{curl}}(\nabla u),$$

or

$$\Pi^{\mathrm{curl}}\left(\nabla(\Pi^{\mathrm{grad}} u - u)\right) = 0.$$

Consequently, it is sufficient to show that the $H(\mathrm{curl})$ d.o.f. applied to $\nabla(\Pi^{\mathrm{grad}} u - u)$ are zero. Consider edge e_{ij} connecting vertex a_i with vertex a_j. Set $E = \nabla(\Pi^{\mathrm{grad}} u - u)$. Then

$$\frac{1}{|a_j - a_i|} \int_{e_{ij}} (\nabla(\Pi^{\mathrm{grad}} u - u)) \cdot (a_j - a_i)$$

$$= \frac{1}{|a_j - a_i|} \int_0^1 \nabla(\Pi^{\mathrm{grad}} u - u)(a_i + t(a_j - a_i)) \cdot (a_j - a_i) |a_j - a_i| \, dt$$

$$= \int_0^1 \frac{d}{dt}(\Pi^{\mathrm{grad}} u - u)(a_i + t(a_j - a_i)) \, dt$$

$$= (\Pi^{\mathrm{grad}} u - u)(a_j) - (\Pi^{\mathrm{grad}} u - u)(a_i) = 0.$$

We are done.

The second commutativity property reads as follows:

$$\boldsymbol{\nabla} \times (\Pi^{\mathrm{curl}} E) \stackrel{?}{=} \Pi^{\mathrm{div}}(\boldsymbol{\nabla} \times E).$$

Again, by the shape functions reproducibility property, this is equivalent to

$$\Pi^{\mathrm{div}}(\boldsymbol{\nabla} \times (\Pi^{\mathrm{curl}} E - E)) = 0.$$

Vanishing of the interpolant is equivalent to vanishing of all d.o.f., i.e., there must be

$$\int_f \underbrace{\boldsymbol{\nabla} \times (\Pi^{\mathrm{curl}} E - E) \cdot n_f}_{\mathrm{curl}_f(\Pi^{\mathrm{curl}} E - E)} = 0$$

for each face f. But, by the Stokes Theorem, the face integral is equal to

$$\int_{\partial f} (\Pi^{\mathrm{curl}} E - E)_t = \sum_e \int_e (\Pi^{\mathrm{curl}} E - E)_t = 0$$

by the definition of operator Π^{curl}.

Finally, we have the third commutativity property,

$$P(\boldsymbol{\nabla} \cdot v) \stackrel{?}{=} \boldsymbol{\nabla} \cdot (\Pi^{\mathrm{div}} v).$$

By the shape functions reproducibility property, it is equivalent to proving that

$$P(\boldsymbol{\nabla} \cdot (\Pi^{\mathrm{div}} v - v)) = 0,$$

or

$$\int_K \boldsymbol{\nabla} \cdot (\Pi^{\mathrm{div}} v - v) = 0.$$

But this follows immediately from the Gauss Theorem and the definition of operator Π^{div},

$$\int_K \boldsymbol{\nabla} \cdot (\Pi^{\mathrm{div}} v - v) = \int_{\partial K} (\Pi^{\mathrm{div}} v - v) \cdot n = \sum_f \int_f (\Pi^{\mathrm{div}} v - v) \cdot n_f = 0. \qquad \square$$

Lowest order hexahedral element of the first type. Let $K = [a_1, b_1] \times [a_2, b_2] \times [a_3, b_3]$. The choice of spaces is perhaps now more natural as it is simply implied by examining a range of grad, curl, and div operators. We have

$$W^1 = \mathcal{Q}^{(1,1,1)} := \mathcal{P}^1 \otimes \mathcal{P}^1 \otimes \mathcal{P}^1,$$

$$Q^1 = \mathcal{Q}^{(0,1,1)} \times \mathcal{Q}^{(1,0,1)} \times \mathcal{Q}^{(1,1,0)},$$

$$V^1 = \mathcal{Q}^{(1,0,0)} \times \mathcal{Q}^{(0,1,0)} \times \mathcal{Q}^{(0,0,1)},$$

$$Y^1 = \mathcal{Q}^{(0,0,0)}.$$

Make a quick count to see that the dimensions of the spaces match the number of vertices, edges, and faces. This is consistent with the fact that tangential components of fields from Q^1 and normal components of fields from V^1 are constant along the edges and over faces, respectively.

We can use exactly the same d.o.f. as for the tetrahedral element. Characterization of interpolation operators (3.2.11) and Theorem 3.3 (including the structure of the proof) remain valid for the hexahedral element as well.

Shape functions for the lowest order hexahedron are defined as the tensor product of 1D affine coordinates λ_i, μ_i, ν_i, $i = 0, 1$, corresponding to the three directions. We start with H^1 vertex shape functions:

$$\lambda_i(x_1)\mu_j(x_2)\nu_k(x_3),$$

$$(i, j, k) = (0, 0, 0), (1, 0, 0), (0, 1, 0), (1, 1, 0), (0, 0, 1), (1, 0, 1), (0, 1, 1), (1, 1, 1),$$

where we use the lexicographic ordering for the vertices.

The $H(\text{curl})$ shape functions are implied by the grad operator. For the four edges parallel to the x_1 axes, we have

$$(\lambda_1'(x_1)\mu_j(x_2)\nu_k(x_3), 0, 0), \quad (j, k) = (0, 0), (0, 1), (1, 0), (1, 1),$$

where (j, k) correspond to the vertices in the x_2-x_3 plane. Note that the shape functions are vector-valued with the second and third components vanishing. The tangential component evaluated along one of the four edges is constant and equals either one or zero. In exactly the same way, we define the shape functions corresponding to edges parallel to the x_2 axis, and then those for edges parallel to x_3 axes.

The $H(\text{div})$ shape functions are implied by the action of curl operator. For the two faces normal to the x_1 axis, we have

$$(\lambda_i(x_1)\mu_1'(x_2)\nu_1'(x_3), 0, 0), \quad i = 0, 1.$$

In the same way we define the four remaining shape functions. Finally, the L^2 shape function is just a constant.

The shape functions provide a dual basis to the same d.o.f. as for the lowest order tetrahedron (with orientations implied by the lexicographic rule); see Exercise 3.2.7. Note that, due to the invariance of 1D affine coordinates with respect to 1D affine isomorphisms, the formulas for the shape functions remain valid for a hexahedron of arbitrary size.

3.2.3 ▪ Right Inverses of Grad, Curl, Div Operators

Let $A : X \to Y$ be a linear operator from a vector space X into a vector space Y. Recall that operator $B : Y \supset \mathcal{R}(A) \to X$ is called a *right inverse* of operator A if

$$ABy = y, \quad y \in \mathcal{R}(A),$$

i.e., composition AB, restricted to the range of A, reduces to identity. The right inverses for grad, curl, and div operators discussed in this section provide very useful tools for studying the exact sequence for both continuous and discrete spaces.

Define

$$(GE)(x) := x \cdot \int_0^1 E(tx)\, dt,$$

$$(Kv)(x) := -x \times \int_0^1 tv(tx)\, dt, \quad (3.2.13)$$

$$(D\psi)(x) := x \int_0^1 t^2 \psi(tx)\, dt$$

or, componentwise,

$$(GE)(x) = x_j \int_0^1 E_j(tx)\, dt\,,$$

$$(Kv)_i(x) = -\epsilon_{ijk} x_j \int_0^1 t v_k(tx)\, dt\,,$$

$$(D\psi)_i(x) = x_i \int_0^1 t^2 \psi(tx)\, dt\,.$$

Operators G, K, D provide *right inverses* for grad, curl, and div operators, respectively. In fact, we have a slightly stronger result:

$$\boldsymbol{\nabla} \times E = 0 \quad \Rightarrow \quad \boldsymbol{\nabla}(GE) = E\,,$$

$$\boldsymbol{\nabla} \cdot v = 0 \quad \Rightarrow \quad \boldsymbol{\nabla} \times (Kv) = v\,,$$

$$\boldsymbol{\nabla} \cdot D\psi = \psi\,.$$

The identities follow immediately from a more general result relating the three operators.

Lemma 3.4. *The following identities hold:*

$$\boldsymbol{\nabla} \cdot D\psi = \psi\,,$$

$$\boldsymbol{\nabla} \times Kv = v - D(\boldsymbol{\nabla} \cdot v)\,, \tag{3.2.14}$$

$$\boldsymbol{\nabla}\, GE = E - K(\boldsymbol{\nabla} \times E)$$

for sufficiently regular scalar-valued function ψ and vector-valued functions v, E.

Proof. The proofs rely on elementary computations and $\epsilon - \delta$ identity (see Exercise 3.2.2). We start with

$$(D\psi)_i(x) := x_i \int_0^1 t^2 \psi(\underbrace{tx}_{=y})\, dt\,.$$

Then

$$\frac{\partial}{\partial x_i}(D\psi)_i = 3\int_0^1 t^2 \psi(tx)\, dt + x_i \int_0^1 t^2 \frac{\partial \psi}{\partial y_j} t\delta_{ji}\, dt$$

$$= \int_0^1 \frac{d}{dt}(t^3 \psi(tx))\, dt$$

$$= t^3 \psi(tx)\,\big|_0^1 = \psi(x)\,.$$

Similarly,

$$\epsilon_{ijk}\frac{\partial}{\partial x_j}\left(-\epsilon_{klm}\, x_l \int_0^1 t v_m(tx)\, dt\right)$$

$$= -\left[(\delta_{il}\delta_{jm} - \delta_{im}\delta_{jl})\left(\delta_{lj}\int_0^1 t v_m(tx)\, dt + x_l \int_0^1 t\frac{\partial v_m}{\partial y_j} t\, dt\right)\right]$$

$$= -\int_0^1 t v_i(tx)\, dt + 3\int_0^1 t v_i(tx)\, dt - x_i \int_0^1 t\frac{\partial v_j}{\partial y_j} t\, dt + x_j \int_0^1 t\frac{\partial v_i}{\partial y_j} t\, dt$$

$$= \int_0^1 \frac{d}{dt}[t^2 v_i(tx)]\, dt - x_i \int_0^1 t^2 \frac{\partial v_j}{\partial y_j}(tx)\, dt$$

$$= v_i(x) - x_i \int_0^1 t^2 \frac{\partial v_j}{\partial y_j}(tx)\, dt\,.$$

The last identity is perhaps the most difficult to prove. We will start with the $K(\nabla \times E)$ term.

$$-(K\nabla \times E)_i = \epsilon_{ijk} x_j \int_0^1 t\epsilon_{klm} \frac{\partial E_m}{\partial y_l}(tx)\, dt$$

$$= (\delta_{il}\delta_{jm} - \delta_{im}\delta_{lj})x_j \int_0^1 t\frac{\partial E_m}{\partial y_l}(tx)\, dt$$

$$= x_j \int_0^1 t\frac{\partial E_j}{\partial y_i}(tx)\, dt - x_j \int_0^1 t\frac{\partial E_i}{\partial y_j}(tx)\, dt\,.$$

The second term in the last line is equal to

$$-\int_0^1 \frac{d}{dt}[tE_i(tx)]\, dt + \int_0^1 E_i(tx)\, dt = E_i(x) + \int_0^1 E_i(tx)\,.$$

On the other side,

$$\frac{\partial}{\partial x_i}\left[x_j \int_0^1 E_j(tx)\, dt\right] = \delta_{ij} \int_0^1 E_j(tx)\, dt + x_j \int_0^1 t\frac{\partial E_j}{\partial y_i}(tx)\, dt\,.$$

Compare the terms to finish the proof.   □

3.2.4 ▪ Elements of Arbitrary Order

Hexahedral element of arbitrary order of the first type.   The reasoning behind the construction of the lowest order hexahedron extends easily to hexahedra of arbitrary and *anisotropic* polynomial order. We introduce the following spaces:

$$W^p = \mathcal{P}^p \otimes \mathcal{P}^q \otimes \mathcal{P}^r\,,$$
$$Q^p = (\mathcal{P}^{p-1} \otimes \mathcal{P}^q \otimes \mathcal{P}^r) \times (\mathcal{P}^p \otimes \mathcal{P}^{q-1} \otimes \mathcal{P}^r) \times (\mathcal{P}^p \otimes \mathcal{P}^q \otimes \mathcal{P}^{r-1})\,,$$
$$V^p = (\mathcal{P}^p \otimes \mathcal{P}^{q-1} \otimes \mathcal{P}^{r-1}) \times (\mathcal{P}^{p-1} \otimes \mathcal{P}^q \otimes \mathcal{P}^{r-1}) \times (\mathcal{P}^{p-1} \otimes \mathcal{P}^{q-1} \otimes \mathcal{P}^r)\,,$$
$$Y^p = \mathcal{P}^{p-1} \otimes \mathcal{P}^{q-1} \otimes \mathcal{P}^{r-1}\,,$$

or, using Ciarlet notation for tensor products, $\mathcal{Q}^{(p,q,r)} := \mathcal{P}^p \otimes \mathcal{P}^q \otimes \mathcal{P}^r$,

$$W^p = \mathcal{Q}^{(p,q,r)}\,,$$
$$Q^p = \mathcal{Q}^{(p-1,q,r)} \times \mathcal{Q}^{(p,q-1,r)} \times \mathcal{Q}^{(p,q,r-1)}\,,$$
$$V^p = \mathcal{Q}^{(p,q-1,r-1)} \times \mathcal{Q}^{(p-1,q,r-1)} \times \mathcal{Q}^{(p-1,q-1,r)}\,,$$
$$Y^p = \mathcal{Q}^{(p-1,q-1,r-1)}\,.$$

Note that the tensor product element allows for a different order of approximation in each direction. The hexahedra are naturally *anisotropic* as opposed to the tetrahedral elements discussed next, which are *isotropic*.

Tetrahedral element of arbitrary order of the second type.

$$W^p = \mathcal{P}^p\,,$$
$$Q^p = \mathcal{P}^{p-1} \times \mathcal{P}^{p-1} \times \mathcal{P}^{p-1}\,,$$
$$V^p = \mathcal{P}^{p-2} \times \mathcal{P}^{p-2} \times \mathcal{P}^{p-2}\,,$$
$$Y^p = \mathcal{P}^{p-3}\,.$$

The choice of spaces reflects a simple fact that, with each differentiation, the polynomial degree goes down by one. Notice that for the hexahedral element, the polynomial order went down by one (in all directions) only at the end of the sequence. This makes these two families of elements incompatible in hybrid meshes, and it is natural to look for another choice of spaces for the tetrahedral element of arbitrary order. The $H(\mathrm{curl})$ elements were introduced by Jean Claude Nédélec in his two fundamental papers published in 1980 and 1986 [62, 63]. Elements of the "first type" were introduced in the first and of "second type" in the second paper. The hexahedral $H(\mathrm{curl})$ element of the second type has no corresponding exact sequence family behind it,[7] and we do not present it here.

Tetrahedral element of arbitrary order of the first type. We introduce the following spaces:

$$W^p = \mathcal{P}^p\,,$$

$$Q^p = (\mathcal{P}^{p-1} \times \mathcal{P}^{p-1} \times \mathcal{P}^{p-1}) \oplus \mathcal{N}^p\,,$$

$$V^p = (\mathcal{P}^{p-1} \times \mathcal{P}^{p-1} \times \mathcal{P}^{p-1}) \oplus \mathcal{RT}^p\,,$$

$$Y^p = \mathcal{P}^{p-1}\,,$$

where

$$\mathcal{N}^p := \{E \in \tilde{\mathcal{P}}^p \times \tilde{\mathcal{P}}^p \times \tilde{\mathcal{P}}^p : x \cdot E(x) = 0 \quad \forall x\}\,,$$

$$\mathcal{RT}^p := \{\phi(x)\, x = \phi(x)(x_1, x_2, x_3) : \phi \in \tilde{\mathcal{P}}^{p-1}\}$$

with $\tilde{\mathcal{P}}^p$ denoting scalar-valued *homogeneous polynomials* of order p. Notation honors the inventors of the elements, Pierre-Armaud Raviart and Jean-Marie Thomas, who introduced $H(\mathrm{div})$-conforming elements [68], and Jean-Claude Nédélec, who introduced $H(\mathrm{curl})$-conforming elements [62, 63].

Note that the polynomial order drops now only by one at the end of the sequence. Construction behind this element is much more subtle than for the tetrahedron of the second type. Taking the gradient of polynomials from $\mathcal{P}^p$, we obtain polynomials in $(\mathcal{P}^{p-1})^3$. We do not accept them for space Q^p though. Instead we complement them with additional polynomials of order p in such a way that (a) $\mathrm{curl}\, Q^p$ will contain complete polynomials of order $p-1$, and (b) we keep the exact sequence structure, i.e., the extra polynomials *do not contain* gradients. This philosophy is already present in the construction of the tetrahedron of the lowest order although its construction is also driven very much by geometry (number of edges and faces). The Nédélec spaces can be introduced and characterized in many different ways, none of them trivial; see, e.g., [25, 35]. In these notes, we will limit ourselves to proving that the spaces form indeed an exact sequence. Indeed, consider the differential complex for the tetrahedron of the first type,

$$\mathcal{P}^p \xrightarrow{\nabla} (\mathcal{P}^{p-1})^3 \oplus \mathcal{N}^p \xrightarrow{\nabla \times} (\mathcal{P}^{p-1})^3 \oplus \mathcal{RT}^p \xrightarrow{\nabla \cdot} \mathcal{P}^{p-1}\,.$$

As the differentiation lowers the (total) polynomial degree by one, the sequence is well-defined and it automatically inherits the structure of the differential complex, i.e., the range of each operator is in the null space of the next operator in the sequence.

In the proof of the exactness of the sequence, the right inverses of grad, curl, div operators come in handy. Let $E = E_{p-1} + \tilde{E}_p$ where $E_{p-1} \in (\mathcal{P}^{p-1})^3$ and $\tilde{E}_p \in \mathcal{N}^p$. Assume that

---

[7]Lacking the exact sequence property negatively impacts stability properties of $H(\mathrm{curl})$-conforming discretizations, an issue that was not understood in 1986.

$\nabla \times E = 0$. According to the right inverse formula,

$$E_{p-1} + \tilde{E}_p = \nabla(GE_{p-1} + \underbrace{G\tilde{E}_p)}_{=0}) = \nabla(GE_{p-1}) = E_{p-1},$$

which proves that $\tilde{E}_p = 0$ and $E = E_{p-1}$ is the gradient of $GE_{p-1} \in \mathcal{P}^p$.

Similarly, let $v = v_{p-1} + \tilde{v}_p$ where $v_{p-1} \in (\mathcal{P}^{p-1})^3$ and $\tilde{v}_p \in \mathcal{RT}^p$. Assume that $\nabla \cdot v = 0$. By the right inverse formula then,

$$v_{p-1} + \tilde{v}_p = \nabla \times (Kv_{p-1} + \underbrace{K\tilde{v}_p)}_{=0} = \nabla \times (Kv_{p-1}) = v_{p-1}.$$

Function Kv_{p-1} is obviously in $(\mathcal{P}^p)^3$, but we need to argue yet that it actually lives in the *smaller* subspace defining $Q^p := (\mathcal{P}^{p-1})^3 \oplus \mathcal{N}^p$. To see it, it is sufficient to decompose v_{p-1} into a polynomial of order $p-2$ and a homogeneous polynomial of order $p-1$; definition of map K then implies that

$$v_{p-1} = v_{p-2} + \tilde{v}_{p-1} \qquad \Rightarrow \qquad Kv_{p-1} = \underbrace{Kv_{p-2}}_{\in(\mathcal{P}^{p-1})^3} + \underbrace{K\tilde{v}_{p-1}}_{\in\mathcal{N}^p}.$$

A similar argument is used to prove the surjectivity of the div operator. Let $\psi_{p-1} \in Y^p = \mathcal{P}^{p-1}$. We have

$$\psi_{p-1} = \psi_{p-2} + \tilde{\psi}_{p-1} \qquad \Rightarrow \qquad D\psi_{p-1} = \underbrace{D\psi_{p-2}}_{\in(\mathcal{P}^{p-1})^3} + \underbrace{D\tilde{\psi}_{p-1}}_{\in\mathcal{RT}^p}.$$

A similar argument can be repeated for the (easier) case of tetrahedron of the second type and hexahedron of the first type; compare Exercise 3.2.3.

3.2.5 ▪ Elements of Variable Order

All discussed elements and analogous, nondiscussed, constructions of prismatic and pyramid elements can be generalized to the case of *elements of variable order*. Let us start with the discussion of the 2D exact sequence:

$$W^p \xrightarrow{\nabla} Q^p \xrightarrow{\text{curl}} Y^p.$$

Square element of the first type of variable order.   The standard element spaces are as follows:

$$W^p = Q^{(p,q)},$$
$$Q^p = Q^{(p-1,q)} \times Q^{(p,q-1)},$$
$$Y^p = Q^{(p-1,q-1)}.$$

Concept of hierarchical shape functions suggests that instead of thinking about the polynomial order for the whole element, we can identify separate orders for the element edges and the element interior. Fig. 3.3 illustrates the concept of the master square element of variable order. Each of the four element edges is assigned a (possibly) different order of approximation: p_1, p_2, q_1, q_2, with anisotropic element (interior) order being (p, q). We request the satisfaction of the *minimum rule*:

$$p_1, p_2 \le p \quad \text{and} \quad q_1, q_2 \le q.$$

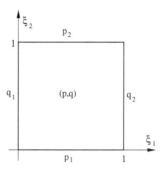

Figure 3.3: Master square element of variable order.

The element energy spaces are defined now as follows:

$$W^p := \{u \in Q^{(p,q)} \; : \qquad\qquad u(\cdot, 0) \in \mathcal{P}^{p_1}(0,1) \,, u(\cdot, 1) \in \mathcal{P}^{p_2}(0,1) \,,$$
$$u(0, \cdot) \in \mathcal{P}^{q_1}(0,1) \,, u(1, \cdot) \in \mathcal{P}^{q_2}(0,1)\} \,,$$
$$Q^p := \{E \in Q^{(p-1,q)} \times Q^{(p,q-1)} \; : \; E_t(\cdot, 0) \in \mathcal{P}^{p_1-1}(0,1) \,, E_t(\cdot, 1) \in \mathcal{P}^{p_2-1}(0,1) \,,$$
$$E_t(0, \cdot) \in \mathcal{P}^{q_1-1}(0,1) \,, E_t(1, \cdot) \in \mathcal{P}^{q_2-1}(0,1)\} \,,$$
$$Y^p := Q^{(p-1,q-1)} \,.$$

$$(3.2.15)$$

One can show that the spaces form an exact sequence; compare Exercise 3.2.10.

Triangle of the first type of variable order. With each edge e of the triangle, we associate a (possibly) different order p_e requesting the minimum rule,

$$p_e \leq p \,, \quad e = 1,2,3 \,.$$

The element energy spaces are defined now as follows:

$$W^p := \{u \in \mathcal{P}^p \; : \; u\big|_e \in \mathcal{P}^{p_e}(e), \, e = 1,2,3\} \,,$$
$$Q^p := \{E \in \mathcal{P}^{p-1} \oplus \mathcal{N}^p \; : \; E_t\big|_e \in \mathcal{P}^{p_e-1}(e) \,, e = 1,2,3\} \,, \qquad (3.2.16)$$
$$Y^p := \mathcal{P}^{p-1} \,.$$

Again, one can show that the spaces form an exact sequence; compare Exercise 3.2.10. We can use now hybrid meshes consisting of square and triangular elements of different order as long as we satisfy the *minimum rule*, i.e., the order for each edge in the mesh is set to the minimum of the orders of neighboring elements (accounting for the anisotropy of square element). Note that the use of the minimum rule implies that all polynomial spaces are well-defined. In particular, once the spaces are specified, we can analyze convergence without discussing choice of shape functions. As for meshes of elements with uniform order, the FE solution depends upon the choice of spaces only.

By now, you should have grasped the idea of variable order elements. The definitions for 3D elements are exactly the same with (possibly) different polynomial orders assigned to edges, faces, and elements. Restrictions of 3D polynomial shape functions to faces form 2D exact sequences on faces, and their restrictions to edges form 1D polynomial sequences on edges. Conformity requires that any two adjacent elements share a 2D sequence on the common face. Similarly, all elements adjacent to a common edge must share a 1D sequence on the edge.

3.2.6 ▪ Shape Functions

As for H^1-conforming elements, shape functions for the remaining energy spaces can be introduced by defining first d.o.f., or directly, by providing bases for the discrete energy spaces. We
shall follow the second route by discussing the shape functions first and delegating construction
of interpolation operators to Section 3.3. In the following discussion, we will stick with elements
forming the first Nédélec family.

Topological classification of shape and basis functions. We have already learned that
the H^1-conforming shape and basis functions can be naturally classified into *vertex, edge, face,*
and *element interior* shape (basis) functions. More precisely, the edge and face shape functions
correspond to the interiors of edges and faces. We call them *edge, face, and element bubbles.*
One can identify their common topological properties without referring to a specific construction.
Each vertex has just one corresponding basis function—the union of vertex shape functions for
all elements adjacent to the vertex (extended by zero to the rest of the mesh). In view of our
discussion on variable order elements, it is natural to assign a separate order of approximation
p_e for each edge e in the mesh. There is then precisely $p_e - 1$ basis functions associated with
the edge. These basis functions are unions of element edge shape functions over all elements
sharing the edge. The restriction of edge basis functions (element edge shape functions) to the
edge coincides with a 1D H^1 bubble, hence the name of "edge bubbles." Extension of a 1D edge
bubble into neighboring faces and then elements is consistent with the definition of face and
element spaces. An edge bubble of order p extends into a polynomial of order p on each adjacent
triangular face and a tensor product of order $(p, 1)$ or $(1, p)$ on each adjacent rectangular face.
The face extensions are then extended into neighboring elements using polynomials from the
element spaces. For a tetrahedron, we use polynomials of order p, and for a hexahedron, we
use tensor products of order $(p, 1, 1)$ or $(1, p, 1)$ or $(1, 1, p)$. Analogous extensions are used for
prisms and pyramids. The most complicated pyramid element space of shape functions includes
also nonpolynomial shape functions.

Similarly, face bubbles start with 2D H^1 bubbles defined on a triangle or a rectangle. For a
triangular face of order p_f, we have exactly $(p_f - 2)(p_f - 1)/2$ bubbles, and for a rectangular
face of order (p_f, q_f), we have $(p_f - 1)(q_f - 1)$ bubbles. These face bubbles are then extended
into neighboring elements.[8] Extension into hexahedral elements will be of order $(p_f, q_f, 1)$ (or a
permutation of it), and extension from a triangle to a tetrahedron will be of order p_f. The support
of a face bubble basis function spans thus at most two elements. Finally, an element bubble basis
function coincides with one of the element bubbles extended by zero to the rest of the mesh. The
discussed topological properties of edge, face, and element bubbles are universal and apply to all
specific constructions of H^1 shape and basis functions.

The topological structure behind H^1 basis and shape functions continues throughout the rest
of the exact sequence. The $H(\text{curl})$-conforming basis and shape functions classify into edge,
face, and element bubbles. Note that there are no vertex shape functions in this group. Similarly,
$H(\text{div})$-conforming basis and shape functions contain face and element bubbles but there are
neither vertex nor edge shape functions in that group. And, finally, the L^2-conforming shape
functions include only element bubbles.

Assembly and orientation-embedded shape functions. Each of the topological entities,
an edge, face, or element interior, comes with its own coordinate that defines the global orientation of the entity. In the standard implementation, element shape functions are defined in the
element system of coordinates disregarding the global edge or face orientations. The global edge

---

[8]Two for an interior face, and just one for a face on the boundary.

Entity	order	H^1	$H(\mathrm{curl})$	$H(\mathrm{div})$	L^2
vertex	1	1	-	-	-
edge	p	$p-1$	p	-	-
trian face	p	$\frac{1}{2}(p-2)(p-1)$	$(p-1)p$	$\frac{1}{2}p(p+1)$	-
recta face	(p,q)	$(p-1)(q-1)$	$p(q-1)+(p-1)q$	pq	-
tet	p	$\frac{1}{6}(p-3)(p-2)(p-1)$	$\frac{1}{2}(p-2)(p-1)p$	$\frac{1}{2}(p-1)p(p+1)$	$\frac{1}{6}p(p+1)(p+2)$
hexa	(p,q,r)	$(p-1)(q-1)(r-1)$	$p(q-1)(r-1)+(p-1)q(r-1)+(p-1)(q-1)r$	$(p-1)qr+p(q-1)r+pq(r-1)$	pqr
prism	(p,q)	$\frac{1}{2}(p-2)(p-1)(q-1)$	$(p-1)p(q-1)+\frac{1}{2}(p-2)(p-1)q$	$\frac{1}{2}p(p+1)q+(p-1)p(q-1)$	$\frac{1}{2}p(p+1)q$
pyramid	p	$(p-1)^3$	$3(p-1)^2p$	$3(p-1)p^2$	p^3

Table 3.1: Number of bubbles for vertices, edges, faces, and element interiors for different energy spaces.

or face orientations must be accounted for during the assembly procedure mirroring the definition of basis functions in terms of shape functions. For Lagrange elements this reduces to the change of enumeration of d.o.f. (shape functions) during the assembly procedure. For hierarchical shape functions, the definition of global basis functions involves additionally sign factors that have to be accounted for during the assembly procedure. In the case of triangular faces there is a head-on conflict between the use of arbitrary systems of coordinates for elements and the hierarchical shape functions which results in the necessity of setting up the element systems of coordinates in a special way; see [35] for a detailed discussion. A great simplification comes from the concept of orientation-embedded shape functions used in [47]. Instead of using predefined shape functions in an element system of coordinates, we define edge and face bubbles in global edge and face coordinates[9] and extend them in an appropriate way into the adjacent elements. This means that we have to communicate the global edge and face orientations to the element shape functions routine[10] and define "on the fly" the element edge and shape function accounting for the orientations. If the shape functions are defined in terms of affine coordinates or products of such, this is accomplished by swapping different coordinates with each other; see [47] for details.

Use of Legendre and Jacobi polynomials. The FE solution depends exclusively upon the FE spaces only in the case of perfect arithmetic. In practice, the round-off error depends strongly upon the specific construction of shape functions, which explains the large number of publications devoted to the construction of different shape functions for the same element spaces. Intuitively, controlling the condition number of element stiffness and mass matrices translates into enforcing (limited) orthogonality in appropriate energy inner products and leads to the use of special functions: Legendre and Jacobi polynomials and their integrals; see again [47] for a literature review and discussion on the subject. As the analysis tools presented in this monograph do not account for the round-off error, we will not discuss these constructions here.

3.2.7 ▪ Parametric Elements and Piola Transforms (Pullback Maps)

The idea of a parametric element can be generalized to the remaining exact sequence energy spaces. Given the exact sequence for a master element $\hat{K}$, we seek transforms (pullbacks) for the

---

[9]Edges and faces "own" the corresponding basis functions.

[10]With respect to the element system of coordinates.

energy spaces defined over an arbitrary (possibly curvilinear) *physical element* K that will make the following diagram commute:

$$
\begin{array}{ccccccc}
H^1(\hat{K}) & \stackrel{\hat{\nabla}}{\longrightarrow} & H(\mathrm{curl}, \hat{K}) & \stackrel{\hat{\nabla}\times}{\longrightarrow} & H(\mathrm{div}, \hat{K}) & \stackrel{\hat{\nabla}\cdot}{\longrightarrow} & L^2(\hat{K}) \\
\downarrow T^{\mathrm{grad}} & & \downarrow T^{\mathrm{curl}} & & \downarrow T^{\mathrm{div}} & & \downarrow T \\
H^1(K) & \stackrel{\nabla}{\longrightarrow} & H(\mathrm{curl}, K) & \stackrel{\nabla\times}{\longrightarrow} & H(\mathrm{div}, K) & \stackrel{\nabla\cdot}{\longrightarrow} & L^2(K) \, .
\end{array}
\tag{3.2.17}
$$

A general (physical) element K is the image of master element $\hat{K}$ through an *element map*,

$$
x_K \; : \; \hat{K} \ni \xi \to x = x_K(\xi) \in K \, ,
$$

that we assume to be a $C^1(\overline{\hat{K}})$-diffeomorphism, i.e., the map is a bijection, and derivatives of both x_K and its inverse x_K^{-1} exist and are continuous up to the boundary. The first T^{grad} map has already been defined,

$$
T^{\mathrm{grad}} \; : \; H^1(\hat{K}) \ni \hat{u} \to u \in H^1(K), \quad u(x) := \hat{u}(x_K^{-1}(x)) \quad \text{or} \quad u = \hat{u} \circ x_K^{-1} \quad \text{or} \quad \hat{u} = u \circ x_K \, .
$$

Using engineering notation,

$$
u(x) = \hat{u}(\xi(x)) \quad \text{or} \quad \hat{u}(\xi) = u(x(\xi)) \, .
$$

Definition of the remaining maps is a consequence of the commutativity of the pullback maps. The transformation T^{curl} must apply in particular to gradients so we can find it out by computing ∇u,

$$
\frac{\partial u}{\partial x_j} = \frac{\partial \hat{u}}{\partial \xi_i} \frac{\partial \xi_i}{\partial x_j} \, .
$$

This leads to the transform for the $H(\mathrm{curl})$ space:

$$
E_j(x) = \hat{E}_i(\xi(x))) \frac{\partial \xi_i}{\partial x_j}(x) \quad \text{or} \quad E = J^{-T} \hat{E} \circ x_K^{-1} \, ,
$$

where $J = \frac{\partial x_i}{\partial \xi_j}$ denotes the Jacobian matrix of the element map. The objects with hats are always functions of ξ and the objects without hats depend upon x. This leads to the simplified notation

$$
T^{\mathrm{curl}} \; : \; H(\mathrm{curl}, \hat{K}) \ni \hat{E} \to E \in H(\mathrm{curl}, K), \quad \text{where} \quad E = J^{-T} \hat{E} \, .
$$

It goes without saying that the right-hand side must be composed with x_K^{-1} or the left-hand side must be composed with x_K.

The next transformation is determined by computing $\mathrm{curl}\, E$.

$$
\begin{aligned}
(\mathrm{curl}\, E)_i &= \epsilon_{ijk} \frac{\partial E_k}{\partial x_j} = \epsilon_{ijk} \frac{\partial}{\partial x_j} \left(\hat{E}_l \frac{\partial \xi_l}{\partial x_k} \right) \\
&= \epsilon_{ijk} \frac{\partial \hat{E}_l}{\partial x_j} \frac{\partial \xi_l}{\partial x_k} + \underbrace{\epsilon_{ijk} \hat{E}_l \frac{\partial^2 \xi_l}{\partial x_j \partial x_k}}_{=0} \qquad \left(\begin{array}{l} \text{product of a symmetric and} \\ \text{an antisymmetric matrix} = 0 \end{array} \right) \\
&= \epsilon_{ijk} \frac{\partial \hat{E}_l}{\partial \xi_m} \frac{\partial \xi_m}{\partial x_j} \frac{\partial \xi_l}{\partial x_k} = (*) \, .
\end{aligned}
$$

Recall now the definition of inverse Jacobian j^{-1} (determinant of inverse Jacobian matrix J^{-1}),

$$
\epsilon_{ijk} \frac{\partial \xi_1}{\partial x_i} \frac{\partial \xi_2}{\partial x_j} \frac{\partial \xi_3}{\partial x_k} = j^{-1} \, ,
$$

or, more generally,

$$\epsilon_{ijk} \frac{\partial \xi_\alpha}{\partial x_i} \frac{\partial \xi_\beta}{\partial x_j} \frac{\partial \xi_\gamma}{\partial x_k} = j^{-1} \epsilon_{\alpha\beta\gamma} \, .$$

Multiplying both sides by $\partial x_l / \partial \xi_\alpha$, we get

$$\epsilon_{ijk} \underbrace{\frac{\partial x_l}{\partial \xi_\alpha} \frac{\partial \xi_\alpha}{\partial x_i}}_{=\delta_{li}} \frac{\partial \xi_\beta}{\partial x_j} \frac{\partial \xi_\gamma}{\partial x_k} = j^{-1} \epsilon_{\alpha\beta\gamma} \frac{\partial x_l}{\partial \xi_\alpha}$$

or

$$\epsilon_{ljk} \frac{\partial \xi_\beta}{\partial x_j} \frac{\partial \xi_\gamma}{\partial x_k} = j^{-1} \epsilon_{\alpha\beta\gamma} \frac{\partial x_l}{\partial \xi_\alpha} \, .$$

In particular, differentiating both sides with respect to x_l, we learn that

$$\epsilon_{\alpha\beta\gamma} \frac{\partial}{\partial x_l} \left(j^{-1} \frac{\partial x_l}{\partial \xi_\alpha} \right) = \epsilon_{ljk} \frac{\partial^2 \xi_\beta}{\partial x_j \partial x_l} \frac{\partial \xi_\gamma}{\partial x_k} + \epsilon_{ljk} \frac{\partial \xi_\beta}{\partial x_j} \frac{\partial \xi_\gamma}{\partial x_k \partial x_l} = 0 \qquad (3.2.18)$$

as the product of a symmetric and an unsymmetric matrix must vanish.

Returning to our computation of curl E, we get

$$(*) = \epsilon_{\alpha m l} j^{-1} \frac{\partial x_i}{\partial \xi_\alpha} \frac{\partial \hat{E}_l}{\partial \xi_m} = j^{-1} \frac{\partial x_i}{\partial \xi_\alpha} \epsilon_{\alpha m l} \frac{\partial \hat{E}_l}{\partial \xi_m} = j^{-1} \frac{\partial x_i}{\partial \xi_\alpha} (\widehat{\mathrm{curl} \hat{E}})_\alpha \, .$$

This leads to the transformation rule for the $H(\mathrm{div})$ fields,

$$T^{\mathrm{div}} : H(\mathrm{div}, \hat{K}) \ni \hat{H} \to H \in H(\mathrm{div}, K) \, , \quad \text{where} \quad H_i = j^{-1} \frac{\partial x_i}{\partial \xi_\alpha} \hat{H}_\alpha \quad \text{or} \quad H = j^{-1} J \hat{H} \, .$$

Finally, we need to compute div H,

$$\mathrm{div}\, H = \frac{\partial H_i}{\partial x_i} = \underbrace{\frac{\partial}{\partial x_i} \left(j^{-1} \frac{\partial x_i}{\partial \xi_\alpha} \right) \hat{H}_\alpha}_{=0} + j^{-1} \frac{\partial x_i}{\partial \xi_\alpha} \frac{\partial \hat{H}_\alpha}{\partial x_i}$$

$$= j^{-1} \frac{\partial x_i}{\partial \xi_\alpha} \frac{\partial \hat{H}_\alpha}{\partial \xi_\beta} \frac{\partial \xi_\beta}{\partial x_i} = j^{-1} \frac{\partial \hat{H}_\alpha}{\partial \xi_\alpha} = j^{-1} \widehat{\mathrm{div} \hat{H}} \, ,$$

where the underbraced term vanishes, which can be seen by selecting β and γ in (3.2.18) in such a way that $\epsilon_{\alpha\beta\gamma} = 1$ (e.g., for $\alpha = 1$, set $\beta = 2, \gamma = 3$).

The last transformation formula for the L^2 fields reads thus as follows:

$$T : L^2(\hat{K}) \ni \hat{f} \to f \in L^2(K) \, , \quad \text{where} \quad f = j^{-1} \hat{f} \, .$$

The pullback map for the $H(\mathrm{div})$ fields is known in mechanics as the Piola transform, which has motivated me to extend this name to all of the transforms.

Note that, with the regularity assumptions made on the element map, all Piola transforms are well-defined, i.e., they preserve the energy spaces. We now make some crucial observations concerning conformity. Begin with a simple observation that the global C^0-continuity of the union of element maps and the continuity of functions $\hat{u}$ in the parametric domain imply the global continuity of the corresponding functions u in the physical domain. If two sufficiently regular functions are continuous along a curve, the corresponding *tangential* derivative must be the same. As the Piola transform T^{curl} was derived by computing the gradients, we expect that the continuity of tangential components of $H(\mathrm{curl})$ fields will be preserved as well. This is indeed the case. Consider a curve in the parametric domain parametrized with

$$\xi_k = \xi_k(t), \quad t \in [0,1].$$

The image of the curve through the element map is naturally parametrized with the composition of the parametrization in the parametric domain and the element map,

$$x_j = x_j(\xi_k(t)), \quad t \in [0,1].$$

Computing the tangent component of $H(\mathrm{curl})$ E field,

$$\frac{\partial x_j}{\partial \xi_k}\frac{\partial \xi_k}{\partial t}E_j = \frac{\partial x_j}{\partial \xi_k}\frac{\partial \xi_k}{\partial t}\hat{E}_i\frac{\partial \xi_i}{\partial x_j} = \frac{\partial \xi_i}{\partial t}\hat{E}_i,$$

we obtain the tangent component of field $\hat{E}$ in the parametric domain. Equivalently,

$$E_t\,ds = \hat{E}_t\,ds_0,$$

where ds, ds_0 stand for the length of the tangent vectors before the normalization. The Piola map preserves tangent components, and the tangential component of E along the curve in the physical domain depends only upon the restriction of the element map to the corresponding curve in the parametric domain. Now comes the main point. If the union of element maps is globally continuous (C^0 continuity is enough), then $H(\mathrm{curl})$-conforming functions in the parametric domain are mapped into $H(\mathrm{curl})$-conforming functions in the physical domain.

A similar result holds for the $H(\mathrm{div})$ fields. We begin again with the formula for the determinant,

$$\epsilon_{ijk}\frac{\partial x_i}{\partial \xi_\alpha}\frac{\partial x_j}{\partial \xi_\beta}\frac{\partial x_k}{\partial \xi_\gamma} = j\,\epsilon_{\alpha\beta\gamma}.$$

Now let $\xi_\beta(s,t)$ be a parametrization of a surface $\hat{S}$ in the parametric domain. Then $x_j(\xi_\beta(s,t))$ provides a parametrization for the corresponding surface S in the physical domain. Multiplying the formula above by $\frac{\partial \xi_\beta}{\partial s}\frac{\partial \xi_\gamma}{\partial t}$, we obtain

$$\epsilon_{ijk}\frac{\partial x_i}{\partial \xi_\alpha}\frac{\partial x_j}{\partial s}\frac{\partial x_k}{\partial t} = \epsilon_{ijk}\frac{\partial x_i}{\partial \xi_\alpha}\frac{\partial x_j}{\partial \xi_\beta}\frac{\partial \xi_\beta}{\partial s}\frac{\partial x_k}{\partial \xi_\gamma}\frac{\partial \xi_\gamma}{\partial t} = j\,\epsilon_{\alpha\beta\gamma}\frac{\partial \xi_\beta}{\partial s}\frac{\partial \xi_\gamma}{\partial t}.$$

As the cross product of two tangent vectors to a surface gives a normal to the surface, we obtain the relation between normal vectors for $\hat{S}$ and the corresponding image surface S,

$$\frac{\partial x_i}{\partial \xi_\alpha}n_i\,dS = j\hat{n}_\alpha\,dS_0,$$

or

$$n_l\,dS = j\frac{\partial \xi_\alpha}{\partial x_l}\hat{n}_\alpha\,dS_0,$$

where $\hat{n}, n$ are now the unit vectors and dS_0, dS denote the length of normal vectors before normalization. This implies now the relation between normal components of $H(\mathrm{div})$ fields in the parametric and physical domains,

$$n_l H_l\,dS = j\frac{\partial \xi_\alpha}{\partial x_l}\hat{n}_\alpha j^{-1}\frac{\partial x_l}{\partial \xi_\beta}\hat{H}_\beta\,dS_0 = \hat{n}_\alpha\hat{H}_\alpha\,dS_0.$$

Consequently, normal components are preserved, which implies that the Piola transform maps $H(\mathrm{div})$-conforming fields in the parametric domain into $H(\mathrm{div})$-conforming fields in the physical domain.

Exercises

3.2.1. Prove unisolvence for the d.o.f. given by (3.2.10) and the lowest order spaces for the tetrahedron (3.2.9). You can restrict yourself to the master tetrahedron. (3 points)

3.2.2. Prove the $\epsilon - \delta$ identity:

$$\epsilon_{ijm}\epsilon_{klm} = \delta_{ik}\delta_{jl} - \delta_{il}\delta_{jk}\,.$$

Hint: With the right geometrical interpretation of the left-hand side and logical interpretation of the right-hand side, you can "see" the identity. (3 points)

3.2.3. Polynomial exact sequences. Prove that the discussed polynomial sequences for the hexahedron of the first type and tetrahedron of the second type are exact. *Hint:* Look up the reasoning for the tetrahedron of the first type in the text. (3 points)

3.2.4. 2D elements. Given the 3D exact polynomial sequences, write out the corresponding two 2D exact polynomial sequences for the square and triangular elements (a total of six sequences) and prove that they are exact. Be concise. (5 points)

3.2.5. Affine coordinates. Prove the following facts about the affine coordinates:

- The affine coordinates are independent of the enumeration of vertices (in the presented construction, we considered vectors $x - a_0, a_i - a_0$, $i = 1, 2, 3$, so it looks like things might depend upon the choice of vertex a_0).

- The *affine coordinates are invariant under affine transformations*: if λ_i are affine coordinates of a point x with respect to vertices a_i then λ_i are also affine coordinates of a point Tx with respect to vertices Ta_i for any bijective affine map T.

- In 2D, the affine coordinates may be interpreted as *area coordinates*. Prove that

$$\lambda_i = \frac{\text{area of } T_i}{\text{area of } T}, \quad i = 0, 1, 2\,,$$

where subtriangles T_i of triangle T are defined in Fig. 3.4. Generalize the result to 3D.

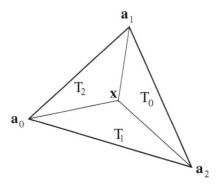

Figure 3.4: Area coordinates.

Be concise. (5 points)

3.2.6. Whitney shape functions (3.2.2). Prove that the Whitney shape functions indeed represent the dual bases corresponding to the d.o.f. specified in the text. *Hint:* Perform the necessary computations in the affine system of coordinates corresponding to $\lambda_j, j = 1, 2, 3$. (5 points)

3.2.7. Shape functions for the lowest order hexahedron. Prove that the shape functions for the lowest order hexahedron provide dual bases to the standard d.o.f. with properly introduced orientations for edges and faces. (3 points)

3.2.8. Characterization of Nédélec's space. Let $\tilde{\mathcal{P}}^k$ denote homogeneous polynomials of order k. Prove the following identity:

$$x \times (\tilde{\mathcal{P}}^{p-1})^3 = \{E \in (\tilde{\mathcal{P}}^p)^3 \ : \ x \cdot E(x) = 0 \quad \forall x\}.$$

(5 points)

3.2.9. Prismatic element. Given the exact sequences for the triangle and the 1D sequence for a unit interval, construct two exact sequences for the prism starting with $W^p = \mathcal{P}^p(T) \otimes \mathcal{P}^q(I)$ where T is a triangle and I an interval. (5 points)

3.2.10. Elements of variable order. Prove that spaces (3.2.15) and (3.2.16) form an exact sequence. (5 points)

3.3 ▪ Projection Based Interpolation

We have introduced the projection based (PB) interpolation in the context of an a posteriori error estimation [65] and generalized it later to the exact sequence spaces in [36]. The name was actually coined by Ralf Hiptmair. I have always claimed that the PB interpolation is unique, provided we accept the following three assumptions to be satisfied by the interpolation operators:

(i) *Locality.* The interpolant in element K should depend upon the values of the interpolated function (and its derivatives) within the same element only.

(ii) *Conformity.* The interpolant should belong to the appropriate energy space, i.e., it should satisfy the corresponding global continuity requirements.

(iii) *Optimality.* Given restrictions resulting from the first two assumptions, the interpolation error should be as small as possible.

The first two assumptions lead to the following observations:

1. The value of the H^1 interpolant at any vertex should coincide with the value of the interpolated function at the same vertex. Indeed, global continuity requires that the vertex value should be the same for all elements sharing the vertex node. On the other side, vertex is the only common part of those elements, so the locality argument leaves no choice—the interpolant value should be set to the function value at the vertex.

2. The H^1 interpolant on an edge should depend only upon the restriction of the interpolated function on the edge. Similarly, the tangential component of an $H(\text{curl})$ interpolant on an edge should depend only upon the tangential component of the interpolated function along the edge.

3. The H^1 interpolant on a face should depend only upon the restriction of the interpolated function to the face. Similarly, the tangential component of an $H(\text{curl})$ interpolant on a face should depend only upon the tangential component of the interpolated function over the face. And, the normal component of an $H(\text{div})$ interpolant on a face should depend only upon the normal component of the interpolated function over the face.

Finally, the optimality criterion leads to local projections: over element edges, faces, and element interiors. The question is, *in what norm or seminorm?* The correct answer[11] comes from the Trace Theorems; we should use fractional norms implied by them. These norms, leading to minimum regularity assumptions, have been analyzed in theory [29, 24, 14, 30, 26], but in practical computations we use stronger integer (and local) norms. This is what we will discuss here. For a recent p error analysis of this version of PB interpolation, see [57]. In these notes, we will restrict ourselves to h estimates only but we will comment later on how p-estimates and the Bramble–Hilbert argument imply the corresponding hp estimates as well.

3.3.1 ▪ PB Interpolation in 1D

Before we review the rather complicated 3D definitions, it will be helpful to discuss first the concept of PB interpolation and the commuting diagram property in one and two space dimensions. Let $I = (a, b)$ be a 1D element. We want to define an H^1-interpolation operator Π^{grad} that will make the following 1D diagram commute:

$$
\begin{array}{ccc}
H^1(I) & \xrightarrow{\ \partial\ } & L^2(I) \\
\downarrow \Pi^{\mathrm{grad}} & & \downarrow P \\
W^p & \xrightarrow{\ \partial\ } & Y^p .
\end{array}
\qquad (3.3.19)
$$

Above, $W^p = \mathcal{P}^p(I)$, $Y^p = \mathcal{P}^{p-1}(I)$, and $P : L^2(I) \to \mathcal{P}^{p-1}(I)$ stands for the L^2-orthogonal projection. The choice of P as an "interpolation operator" for the L^2 energy space is obvious. The space does not involve any global conformity, the L^2-projection onto globally discontinuous polynomials is a local operation, and it is simply the best operator we can have—it delivers the best approximation error in the L^2-norm. For any $y \in L^2(I)$, we have

$$
\int_I (y - Py)\varphi = 0 \quad \forall \varphi \in \mathcal{P}^{p-1}(I) .
$$

Now, the polynomial space $\mathcal{P}^{p-1}(I)$ can be decomposed into two L^2-orthogonal subspaces: constants and polynomials with zero average,

$$
\mathcal{P}^{p-1}(I) = \mathbb{R} \overset{\perp}{\oplus} \mathcal{P}^{p-1}_{\mathrm{avg}}(I) .
$$

The orthogonality condition can thus be written as two independent conditions:

$$
\begin{aligned}
&\int_I (y - Py) = 0 , \\
&\int_I (y - Py)\varphi = 0 \quad \forall \varphi \in \mathcal{P}^{p-1}_{\mathrm{avg}}(I) .
\end{aligned}
\qquad (3.3.20)
$$

Let $\mathcal{P}^p_0(I)$ denote the H^1 element bubbles, i.e., the polynomials of order p that vanish at the endpoints of interval I. It is easy to see that the derivative is a well-defined bijection from $\mathcal{P}^p_0(I)$ onto $\mathcal{P}^{p-1}_{\mathrm{avg}}(I)$,

$$
\partial : \mathcal{P}^p_0(I) \ni u \to u' \in \mathcal{P}^{p-1}_{\mathrm{avg}}(I) .
$$

The second of the orthogonality conditions for the L^2-projection can thus be rewritten in the form

$$
\int_I (y - Py)\varphi' = 0 \quad \forall \varphi \in \mathcal{P}^p_0(I) .
$$

---

[11] Advice of Ivo Babuška.

If we apply the condition to $y = u', u \in H^1(I)$, and use the required commutativity property $P\partial = \partial\Pi^{\text{grad}}$, we obtain a condition defining partially the Π^{grad} operator,

$$\int_I (u - \Pi^{\text{grad}}u)'\varphi' = 0 \quad \forall\, \varphi \in \mathcal{P}_0^p(I)\,. \tag{3.3.21}$$

As discussed earlier, the locality argument forces us to employ the standard interpolation at vertex nodes,

$$(u - \Pi^{\text{grad}}u)(a) = (u - \Pi^{\text{grad}}u)(b) = 0\,. \tag{3.3.22}$$

In conclusion, the Π^{grad} operator is defined by the conditions (3.3.22) and (3.3.21). Equivalently, we can define the operator by

$$H^1(I) \ni u \to \Pi^{\text{grad}}u = u_1 + u_2 \in \mathcal{P}^p(I)\,,$$

where u_1 is the vertex interpolant, and u_2 solves the minimization problem:

$$\int_I \left| \frac{d}{dx}(u - u_1 - u_2) \right|^2 dx \to \min_{u_2 \in \mathcal{P}_0^p(I)}\,. \tag{3.3.23}$$

We finish the proof commutativity by checking condition (3.3.20)$_1$,

$$\int_a^b (u - \Pi^{\text{grad}}u)'\, dx = (u - \Pi^{\text{grad}}u)|_a^b = 0\,.$$

Finally, note that the minimization problem (3.3.23), equivalent to the orthogonality property (3.3.21), is a natural condition to ask on its own as we want to minimize the interpolation error in the H^1-norm. The commutativity property implies that we need to minimize the H^1-seminorm rather than the H^1-norm, Recall that, by the Poincaré lemma, the H^1-norm and seminorms are equivalent over space $H_0^1(I)$.

3.3.2 ▪ PB Interpolation in 2D

We extend now our construction to a 2D element K.

$$\begin{array}{ccccc}
H^1(K) & \xrightarrow{\nabla} & H(\text{curl}, K) & \xrightarrow{\text{curl}} & L^2(K) \\
\downarrow \Pi^{\text{grad}} & & \downarrow \Pi^{\text{curl}} & & \downarrow P \\
W^p & \xrightarrow{\nabla} & Q^p & \xrightarrow{\text{curl}} & Y^p\,.
\end{array} \tag{3.3.24}$$

As in the 1D case, P denotes the L^2-projection. The operators Π^{grad} and Π^{curl} are defined by requesting the following orthogonality properties.

H^1 PB interpolation

$$\begin{cases}
(u - \Pi^{\text{grad}}u)(v) = 0 & \text{for every vertex } v\,, \\[2ex]
\int_e \frac{d}{dt}(u - \Pi^{\text{grad}}u)\frac{\partial\varphi}{\partial t} = 0 & \text{for all edge bubbles } \varphi, \quad \text{for every edge } e\,, \\[2ex]
\int_K \nabla(u - \Pi^{\text{grad}}u)\nabla\varphi = 0 & \text{for all element bubbles } \varphi\,.
\end{cases} \tag{3.3.25}$$

$H(\mathrm{curl})$ PB interpolation

$$\begin{cases} \displaystyle\int_e (E - \Pi^{\mathrm{curl}} E)_t \varphi = 0 & \text{for all edge shape functions } \varphi, \qquad \text{for every edge } e\,, \\[2ex] \displaystyle\int_K \mathrm{curl}(E - \Pi^{\mathrm{curl}} E)\,\mathrm{curl}\,\varphi = 0 & \text{for all element } H(\mathrm{curl}) \text{ bubbles } \varphi\,, \\[2ex] \displaystyle\int_K (E - \Pi^{\mathrm{curl}} E)\boldsymbol{\nabla}\varphi = 0 & \text{for all element } H^1 \text{ bubbles } \varphi\,. \end{cases}$$

$$(3.3.26)$$

Conditions $(3.3.25)_{1,2}$ and $(3.3.26)_1$ correspond to the 1D interpolation applied edgewise on the boundary of the element. Condition $(3.3.25)_3$ is a natural extension of what we did in 1D, and the condition $(3.3.26)_3$ is then implied by the commutativity property. Finally, condition $(3.3.26)_2$ is a result of the commutativity with the L^2-projection. To prove it, we use the same reasoning as in 1D. Space Y^p is decomposed orthogonally into constant functions and functions with zero average,

$$Y^p = \mathbb{R} \overset{\perp}{\oplus} Y^p_{\mathrm{avg}}\,.$$

The L^2-projection is then equivalent to the orthogonality conditions:

$$\int_K (y - Py) = 0\,,$$
$$\int_K (y - Py)\varphi = 0 \quad \forall\,\varphi \in Y^p_{\mathrm{avg}}\,.$$

The commutativity property implies now that

$$\int_K \mathrm{curl}(E - \Pi^{\mathrm{curl}} E) = 0\,,$$
$$\int_K \mathrm{curl}(E - \Pi^{\mathrm{curl}} E)\varphi = 0 \quad \forall\,\varphi \in Y^p_{\mathrm{avg}}\,.$$

Finally, the surjectivity of the curl operator mapping $H(\mathrm{curl})$ element bubbles onto Y^p_{avg} (compare Exercise 3.3.1) implies condition $(3.3.26)_2$. The first condition above is proved in the same way as for the lowest order elements and classical interpolation operators.

Remark 3.5. Surjectivity of the curl operator is part of a bigger picture. Let spaces $W^p(K)$, $Q^p(K), V^p(K), Y^p(K)$ form an exact sequence. Define

$$W^p_0(K) := \{u \in W^p(K) \,:\, u = 0 \text{ on } \partial K\}\,,$$
$$Q^p_0(K) := \{E \in Q^p(K) \,:\, n \times E = 0 \text{ on } \partial K\}\,,$$
$$V^p_0(K) := \{v \in V^p(K) \,:\, n \cdot v = 0 \text{ on } \partial K\}\,,$$
$$Y^p_{\mathrm{avg}}(K) := \left\{y \in Y^p(K) \,:\, \int_K y = 0\right\}\,.$$

Then spaces $W^p_0(K), Q^p_0(K), V^p_0(K), Y^p_{\mathrm{avg}}(K)$ form an exact sequence as well; compare Exercise 3.3.1.

Equivalently, more constructive definitions of the PB interpolants are as follows.

H^1 PB interpolation—equivalent definition

The interpolant is constructed by summing up vertex, edge, and element contributions:

$$H^r(K) \ni u \to \Pi^{\mathrm{grad}} u := u_p = u_1 + u_2 + u_3 \in W^p(K). \qquad (3.3.27)$$

Here

- u_1 is the vertex interpolant constructed using vertex shape functions ϕ_v,

$$u_1(x) := \sum_v u(v)\phi_v(x);$$

- $u_2 := \sum_e u_{2,e}$ is the edge contribution where edge e bubble $u_{2,e}$ is a combination of edge shape functions (edge bubbles),

$$u_{2,e} = \sum_{j=1}^{p-1} u_{2,e}^j \phi_j, \quad \phi_j \in \mathcal{P}_0^p(e),$$

and it is obtained by solving the edge projection problem:

$$\left\| \frac{\partial}{\partial t}(u - (u_1 + u_{2,e})) \right\|_{L^2(e)} \to \min_{u_{2,e} \in \mathcal{P}_0^p(e)};$$

- u_3 is the element bubble obtained by projecting difference $u - u_1 - u_2$ over the element bubbles,

$$\| \nabla(u - (u_1 + u_2 + u_3)) \|_{L^2(K)} \to \min_{u_3}.$$

Above, $\partial/\partial t$ denotes the tangential derivative along the edge. The space of element bubbles depends upon the shape of the element: $\mathcal{P}_0^p(K)$ for a triangle of order p, and $\mathcal{P}_0^p(I) \otimes \mathcal{P}_0^q(I)$ for the master quad $K = I \times I$, $I = (0,1)$.

$H(\mathrm{curl})$ PB interpolation—equivalent definition

The interpolant is defined as the sum of edge and element contributions:

$$H^r(\mathrm{curl}, K) \ni E \to \Pi^{\mathrm{curl}} E := E_p = E_1 + E_2 \in Q^p. \qquad (3.3.28)$$

Here, we have as follows:

- $E_1 = \sum_e E_{1,e}$ is the edge interpolant. Each edge e contribution $E_{1,e}$ lives in the span of edge e shape functions, and it is obtained by solving the L^2 edge projection problem:

$$\| (E - E_{1,e})_t \|_{L^2(e)} \to \min_{E_{1,e}},$$

where E_t denotes the tangential component of vector E. Recall that (the restrictions of) the edge shape functions span $\mathcal{P}^p(e)$.

- E_2 lives in the span of $H(\mathrm{curl})$ element bubbles and is the solution of the constrained L^2-projection problem:

$$\begin{cases} \| \nabla \times (E - E_1 - E_2) \|_{L^2(K)} \to \min_{E_2}, \\ ((E - E_1 - E_2), \nabla \varphi)_{L^2(K)} = 0 \quad \text{for each element } H^1 \text{ bubble } \varphi. \end{cases}$$

The space of $H(\mathrm{curl})$ element bubbles is again defined differently for the triangular and quad elements but the overall procedure defining the operator is independent of the element type. Finally, parameter r above indicates the minimum regularity of interpolated functions for the operations defining the interpolants to be well-defined. The Trace Theorems imply that we should take $r > \frac{1}{2}$. We will discuss it in more detail in Section 3.4.2.

3.3.3 ▪ PB Interpolation in 3D

H^1 PB interpolation

We start with the constructive definition of the operator.

$$H^r(K) \ni u \to \Pi^{\mathrm{grad}} u := u_p = u_1 + u_2 + u_3 + u_4 \in W^p(K), \qquad (3.3.29)$$

where

- u_1 is the vertex interpolant constructed using vertex shape functions ϕ_v,

$$u_1(x) := \sum_v u(v)\phi_v(x);$$

- $u_2 := \sum_e u_{2,e}$ is the edge contribution where edge e bubble $u_{2,e}$ is a combination of edge shape functions (edge bubbles),

$$u_{2,e} = \sum_{j=1}^{p-1} u_{2,e}^j \phi_j, \quad \phi_j \in \mathcal{P}_0^p(e),$$

and it is obtained by solving the edge projection problem,

$$\left\| \frac{\partial}{\partial t}(u - (u_1 + u_{2,e})) \right\|_{L^2(e)} \to \min_{u_{2,e}};$$

- $u_3 := \sum_f u_{3,f}$ is the face contribution where face f bubble $u_{3,f}$ is a combination of face shape functions (face bubbles),

$$u_{3,f} = \sum_j u_{3,f}^j \phi_j,$$

and it is obtained by solving the face projection problem,

$$\|\nabla_t(u - (u_1 + u_2 + u_{3,f}))\|_{L^2(f)} \to \min_{u_{3,f}};$$

- u_4 is the element bubble obtained by projecting difference $u - u_1 - u_2 - u_3$ over the element bubbles,

$$\|\nabla(u - (u_1 + u_2 + u_3 + u_4))\|_{L^2(K)} \to \min_{u_4}.$$

Above, $\partial/\partial t$ denotes the tangential derivative along the edge and ∇_t stands for the tangential component of the gradient. Equivalent variational statements are

$$\int_e \frac{\partial}{\partial t}(u - (u_1 + u_{2,e})) \frac{\partial \varphi}{\partial t} = 0 \qquad \text{for each edge bubble } \varphi,$$

$$\int_f \nabla_t(u - (u_1 + u_2 + u_{3,f})) \cdot \nabla_t \varphi = 0 \qquad \text{for each face bubble } \varphi, \qquad (3.3.30)$$

$$\int_K \nabla(u - (u_1 + u_2 + u_3 + u_4)) \cdot \nabla \varphi = 0 \quad \text{for each element bubble } \varphi.$$

This leads to the equivalent, more compact definition of the interpolant:

$$(u - u_p)(v) = 0 \qquad \text{for each vertex } v\,,$$

$$\int_e \frac{\partial}{\partial t}(u - u_p)\frac{\partial \varphi}{\partial t} = 0 \qquad \text{for each edge bubble } \varphi\,, \qquad \text{for each edge } e\,,$$

$$\int_f \boldsymbol{\nabla}_t(u - u_p) \cdot \boldsymbol{\nabla}_t\varphi = 0 \quad \text{for each face bubble } \varphi\,, \qquad \text{for each face } f\,, \tag{3.3.31}$$

$$\int_K \boldsymbol{\nabla}(u - u_p) \cdot \boldsymbol{\nabla}\varphi = 0 \quad \text{for each element bubble } \varphi\,.$$

$H(\mathrm{curl})$ **PB interpolation**

We start again with the constructive definition of the operator:

$$H^r(\mathrm{curl}, K) \ni E \to \Pi^{\mathrm{curl}} E := E_p = E_1 + E_2 + E_3 \in Q^p\,. \tag{3.3.32}$$

Here we have as follows:

- $E_1 = \sum_e E_{1,e}$ is the edge interpolant. Each edge e contribution $E_{1,e}$ lives in the span of edge e shape functions and it is obtained by solving the edge projection problem:

$$\|(E - E_{1,e})_t\|_{L^2(e)} \to \min_{E_{1,e}}\,,$$

  where E_t denotes the tangential component of vector E.

- $E_2 = \sum_f E_{2,f}$ is the face interpolant, with each face contribution $E_{2,f}$ living in the span of face shape functions (face bubbles) and being the solution of the constrained projection problem:

$$\begin{cases} \|\operatorname{curl}_f(E - E_1 - E_{2,f})\|_{L^2(f)} \to \min_{E_{2,f}}\,, \\ ((E - E_1 - E_{2,f})_t, \boldsymbol{\nabla}_t\varphi)_{L^2(f)} = 0 \quad \text{for each face } H^1 \text{ bubble } \varphi\,. \end{cases}$$

- E_3 lives in the span of element $H(\mathrm{curl})$ bubbles and is the solution of the constrained projection problem:

$$\begin{cases} \|\boldsymbol{\nabla} \times (E - E_1 - E_2 - E_3)\|_{L^2(K)} \to \min_{E_3}\,, \\ ((E - E_1 - E_2 - E_3), \boldsymbol{\nabla}\varphi)_{L^2(K)} = 0 \quad \text{for each element } H^1 \text{ bubble } \varphi\,. \end{cases}$$

Equivalent variational statements lead to the equivalent definition of the interpolant:

$$\int_e (E - E_p)_t\, \psi_t = 0 \qquad\qquad \text{for each edge shape function } \psi\,,$$

$$\qquad\qquad\qquad\qquad\qquad\qquad\qquad \text{for each edge } e\,,$$

$$\int_f \operatorname{curl}_f(E - E_p) \cdot \operatorname{curl}_f \psi = 0 \quad \text{for each } H(\mathrm{curl}) \text{ face bubble } \psi\,,$$

$$\int_f (E - E_p) \cdot \boldsymbol{\nabla}_t\varphi = 0 \qquad\qquad \text{for each } H^1 \text{ face bubble } \varphi\,, \tag{3.3.33}$$

$$\qquad\qquad\qquad\qquad\qquad\qquad\qquad \text{for each face } f\,,$$

$$\int_K \boldsymbol{\nabla} \times (E - E_p) \cdot \boldsymbol{\nabla} \times \psi = 0 \quad \text{for each element } H(\mathrm{curl}) \text{ bubble } \psi\,,$$

$$\int_K (E - E_p) \cdot \boldsymbol{\nabla}\varphi = 0 \qquad\qquad \text{for each element } H^1 \text{ bubble } \varphi\,.$$

$H(\text{div})$ **PB interpolation**

We start again with the constructive definition:

$$H^r(\text{div}, K) \ni v \to \Pi^{\text{div}} v := v_p = v_1 + v_2 \in V^p\,. \tag{3.3.34}$$

Here we have as follows:

- $v_1 = \sum_f v_{1,f}$ is the face interpolant. Each face contribution $v_{1,f}$ lives in the span of the face shape functions, and it solves the projection problem:

$$\|(v - v_{1,f}) \cdot n\|_{L^2(f)} \to \min_{v_{1,f}}\,.$$

- v_2 lives in the span of element $H(\text{div})$ bubbles, and it is the solution of the constrained projection problem:

$$\begin{cases} \|\nabla \cdot (v - v_1 - v_2)\|_{L^2(K)} \to \min_{v,2}\,, \\ ((v - v_1 - v_2), \nabla \times \varphi)_{L^2(K)} = 0 \quad \text{for each element } H(\text{curl}) \text{ bubble } \varphi\,. \end{cases}$$

Equivalent variational statements lead to the equivalent definition of the interpolant:

$$\int_f ((v - v_p) \cdot n)\, \psi \cdot n = 0 \qquad \text{for each } H(\text{div}) \text{ face shape function } \psi\,,$$

$$\qquad\qquad\qquad\qquad\qquad\qquad\qquad\qquad\qquad\qquad \text{for each face } f\,,$$

$$\int_K \nabla \cdot (v - v_p)\, \nabla \cdot \psi = 0 \quad \text{for each element } H(\text{div}) \text{ bubble } \psi\,, \tag{3.3.35}$$

$$\int_K (v - v_p) \cdot \nabla \times \varphi = 0 \qquad \text{for each element } H(\text{curl}) \text{ bubble } \varphi\,.$$

L^2-**projection**

$$L^2(K) \ni f \to Pf := f_p \in Q^p\,, \tag{3.3.36}$$

where

$$\int_K (f - f_p)\, \psi = 0 \quad \text{for each shape function } \psi\,.$$

Theorem 3.6. *Let W^p, Q^p, V^p, Y^p be any FE spaces forming the exact grad-curl-div sequence for any element K. The PB interpolation operators make the de Rham diagram (3.2.12) commute.*

The proof is left to the reader; see Exercises 3.3.6 and 3.3.7.

Remark 3.7. The PB interpolation can be performed on both master and physical elements but the two interpolation operators *do not commute* with the pullback maps. For instance,

$$\widetilde{\Pi^{\text{grad}}u} \neq \hat{\Pi}^{\text{grad}}\hat{u}\,.$$

As the commutativity property is critical for developing interpolation error estimates discussed in the next section, we enforce it by simply interpolating always on the master element, i.e., we pull back the interpolated function u to the master element, interpolate the resulting $\hat{u}$ there, and then push back the interpolant to the physical element. In other words,

$$\widehat{\Pi^{\text{grad}}u} \stackrel{\text{def}}{=} \hat{\Pi}^{\text{grad}}\hat{u}.$$

The same comment applies to the remaining interpolation operators. Note that the resulting interpolation operators commute on the physical element as well.

Exercises

3.3.1. Exact sequence for spaces incorporating homogeneous BCs. Consider spaces defined in Remark 3.5. Prove that the sequence

$$W_0^p(K) \xrightarrow{\nabla} Q_0^p(K) \xrightarrow{\nabla\times} V_0^p(K) \xrightarrow{\nabla\cdot} Y_{\text{avg}}^p(K)$$

is well-defined and exact. (5 points)

3.3.2. H^1 PB interpolation.

(i) Discuss briefly why the three formulations in the text are equivalent.

(ii) Recall the Ciarlet definition of the interpolation operator defined in terms of d.o.f. ψ_j,

$$\Pi u = \sum_{j=1}^{N} \psi_j(u)\phi_j,$$

and prove that it is equivalent to the condition

$$\Pi u \in X(K), \quad \psi_j(u - \Pi u) = 0, \quad j = 1, \dots, N.$$

Here $N = \dim X(K)$ and ϕ_j are the shape functions corresponding to d.o.f. ψ_j.

(iii) Based on characterization (3.3.31), write out the formulas for the d.o.f. corresponding to the PB interpolation.

(iv) Write down explicitly systems of linear equations that need to be solved for computing the edge, face, and interior contributions to the interpolant on a tetrahedral element of order p.

(v) Discuss in a couple of lines why the definition of the PB interpolation holds *for all* H^1-conforming elements including elements of variable order.

(vi) Is the use of hierarchical shape functions necessary for computing the H^1 PB interpolant? Discuss.

(vii) While it is natural to use the shape functions to extend $u_1, u_{2,e}, u_{3,f}$ to the whole element, the final interpolant u_p *is independent* of particular lifts as long as they live in the FE space $X(K) = W_p(K)$. Explain why.

(10 points)

3.3.3. Coding H^1 PB interpolation. The PB interpolant is computed by solving sequentially small systems of linear equations over element edges, faces, and interiors. Suppose you would like to simplify the logic of implementation by solving a single system of linear equations for one element at a time. Try to write down such a system of equations for a 2D triangular element of order p. (3 points)

3.3.4. What are the minimum regularity assumptions for the PB interpolation to be continuous in 3D? In other words, what is the minimum r in (3.3.29)? *Hint:* Recall the trace and Sobolev embedding theorems. (3 points)

3.3.5. $H(\text{curl})$ PB interpolation.

   (i) Write down the variational form of the constrained projection problems. Are the corresponding Lagrange multipliers equal to zero?

   (ii) Following the ideas from Exercise 3.3.2, identify the d.o.f. corresponding to the PB interpolation operator.

   (5 points)

3.3.6. Commutativity of PB interpolation.

   (i) Assume that field E is a gradient, $E = \nabla u$, and prove that so must be the PB interpolant, $E_p = \nabla u_p$, where $u_p \in W^p(K)$.

   (ii) Prove that $u_p = \Pi^{grad} u$. *Hint:* Reduce the definition to the case when both E and E_p are gradients and compare it with the definition of the H^1 interpolant. Recall the discussion for the lowest order Whitney elements.

   (10 points)

3.3.7. Commutativity of PB interpolation (continued). Prove the commutativity of the remaining two blocks in the diagram. (10 points)

3.4 ▪ Classical Interpolation Theory

In this section we develop classical h-interpolation error estimates for the exact sequence energy spaces. For simplicity, we shall restrict ourselves to the sequences of first type only, i.e.,

$$W^p \xrightarrow{\nabla} Q^p \xrightarrow{\nabla \times} V^p \xrightarrow{\nabla \cdot} Y^p,$$

$$\mathcal{P}^p \subset W^p, \quad (\mathcal{P}^{p-1})^N \subset Q^p, \quad (\mathcal{P}^{p-1})^N \subset V^p, \quad \mathcal{P}^{p-1} \subset Y^p.$$

Notice that symbol p in the notation for the space indicates the order of the H^1 element only. The remaining spaces contain complete polynomials of order less than or equal to $p-1$ only.

   In each case, we assume silently that the interpolation operator commutes with the pullback (Piola) transform ("breaking the hat property"), i.e.,

$$\widehat{\Pi u} = \hat{\Pi} \hat{u}.$$

We also assume silently that parameter r specifying the Sobolev regularity of the interpolated function is sufficiently large to ensure the continuity of the interpolation operator.

3.4.1 ▪ Bramble–Hilbert Argument

In the following discussion, $\Omega \subset \mathbb{R}^N, N = 1, 2, \ldots,$ is always a bounded domain. We begin with another version of the Poincaré lemma.

Lemma 3.8. *There exists a positive constant $C = C(\Omega)$ such that*

$$\|u\|^2 \leq C \left\{ \left| \int_\Omega u \right|^2 + \|\nabla u\|^2 \right\} \quad \forall\, u \in H^1(\Omega).$$ (3.4.37)

Proof. See Exercise 3.4.1.　　□

Lemma 3.9. *There exists* $C > 0$ *such that*

$$\|u\|^2_{H^r(\Omega)} \leq C \left\{ \sum_{|\alpha| \leq r-1} \left| \int_\Omega D^\alpha u \right|^2 + |u|^2_{H^r(\Omega)} \right\} \quad \forall\, u \in H^r(\Omega) \tag{3.4.38}$$

for any integer $r > 0$.

Proof. Use Lemma 3.8 and mathematical induction.　　□

Lemma 3.10. *There exists* $C > 0$ *such that*

$$\inf_{\varphi \in \mathcal{P}^{r-1}} \|u - \varphi\|^2_{H^r(\Omega)} \leq C |u|^2_{H^r(\Omega)} \tag{3.4.39}$$

for any integer $r > 0$.

Proof. Apply inequality (3.4.38) to difference $u - \varphi$,

$$\|u - \varphi\|^2_{H^r(\Omega)} \leq C \left\{ \sum_{|\alpha| \leq r-1} \left| \int_\Omega D^\alpha (u - \varphi) \right|^2 + |u|^2_{H^r(\Omega)} \right\}.$$

Note that the r-order derivatives for $r - 1$ order polynomial φ vanish, hence absence of φ in the seminorm on the right-hand side. It remains to show that we can select a polynomial φ in such a way that all averages on the right-hand side vanish. Start by representing φ as the sum of monomials,

$$\varphi = \sum_{|\beta| \leq r-1} c_\beta x^\beta.$$

Now let α be an arbitrary multi-index of order $r - 1$, $|\alpha| = r - 1$. As

$$D^\alpha x^\beta = \begin{cases} \alpha! := \alpha_1! \dots \alpha_n! & \text{if } \alpha = \beta, \\ = 0 & \text{otherwise}, \end{cases}$$

we have

$$D^\alpha \varphi = \sum_{|\beta| \leq r-1} c_\beta D^\alpha x^\beta = \alpha!\, c_\alpha.$$

We can match the constant with the corresponding average of the derivative of function u,

$$|\Omega|\, D^\alpha \phi = |\Omega|\, \alpha!\, c_\alpha = \int_\Omega D^\alpha u.$$

Once all constants c_α, for $|\alpha| = r - 1$, have been selected, we can apply now the same argument to constants corresponding to monomials of one order less,

$$|\Omega|\, D^\alpha x^\alpha = \int_\Omega D^\alpha \left(u - \sum_{|\beta|=r-1} c_\beta x^\beta \right), \quad |\alpha| = r - 2.$$

Proceed by induction to finish the proof.　　□

Corollary 3.11. *Seminorm* $|\cdot|_{H^r(\Omega)}$ *provides an equivalent norm for the quotient space* $H^r(\Omega)$
$/\mathcal{P}^{r-1}$. *In particular, the quotient space equipped with that (semi)norm is complete. Following the same line of argument, we can claim also a more general result for any space of shape functions* W^p *that contains* $\mathcal{P}^{p-1}$. *Replacing* u *with* $u - \varphi$, $\varphi \in W^p$ *in inequality (3.4.39), and taking infimum with respect to* $\varphi \in W^p$ *on both sides, we get*

$$\inf_{\varphi \in W^p} \|u - \varphi\|^2_{H^r(\Omega)} \leq C \inf_{\varphi \in W^p} |u - \varphi|^2_{H^r(\Omega)} \,. \tag{3.4.40}$$

The right-hand side represents thus a norm equivalent to the standard norm in the quotient space $H^r(\Omega)/W^p$.

We arrive at the fundamental result of Bramble and Hilbert.

Theorem 3.12 (Bramble–Hilbert argument for H^r-norm). *Let* Ω *be a domain in* $\mathbb{R}^N$, *and let* W^p *be a subspace of* $H^1(\Omega)$ *such that*

$$\mathcal{P}^p \subset W^p \tag{3.4.41}$$

for some $p \geq 0$. *Let* $r > 0$ *and let* $p + 1 \geq r$. *There exists a constant* $C > 0$, *dependent upon* r, *such that*

$$\inf_{\varphi \in W^p} \|u - \varphi\|_{H^r(\Omega)} \leq C |u|_{H^r(\Omega)} \tag{3.4.42}$$

for every $u \in H^r(\Omega)$.

Proof. Notice that

$$\inf_{\varphi \in W^p} |u - \varphi|_{H^r(\Omega)} \leq |u|_{H^r(\Omega)}$$

and apply inequality (3.4.40).   □

Theorem 3.13 (Bramble–Hilbert argument for $H^r(\mathrm{curl})$-norm). *Let* Ω *be a domain in* $\mathbb{R}^3$ *and let* Q^p *be a subspace of* $H(\mathrm{curl}, \Omega)$ *such that*

$$\mathcal{P}^{p-1} \subset Q^p \quad and \quad \mathcal{P}^{p-1} \subset \boldsymbol{\nabla} \times Q^p \tag{3.4.43}$$

for some $p > 0$. *Let* $r > 0$ *and let* $p \geq r$. *There exists a constant* $C > 0$, *dependent upon* r, *such that*

$$\inf_{\varphi \in Q^p} \left(\|E - \varphi\|^2_{H^r(\Omega)} + \|\boldsymbol{\nabla} \times (E - \varphi)\|^2_{H^r(\Omega)} \right)^{1/2} \leq C \left(|E|^2_{H^r(\Omega)} + |\boldsymbol{\nabla} \times E|^2_{H^r(\Omega)} \right)^{1/2} \tag{3.4.44}$$

for every $E \in H^r(\Omega)$ *such that* $\boldsymbol{\nabla} \times E \in H^r(\Omega)$.

Proof. It is sufficient to prove the result for $p = r$. Consider the space

$$H^r(\mathrm{curl}, \Omega) := \{ E \in H^r(\Omega) \ : \ \boldsymbol{\nabla} \times E \in H^r(\Omega) \} \tag{3.4.45}$$

and the corresponding quotient space:

$$H^r(\mathrm{curl}, \Omega)/Q^p \,. \tag{3.4.46}$$

We have

$$\inf_{\varphi \in Q^p} \left(|E - \varphi|^2_{H^r(\Omega)} + |\boldsymbol{\nabla} \times (E - \varphi)|^2_{H^r(\Omega)} \right)^{1/2}$$
$$\leq \inf_{\varphi \in Q^p} \left(\|E - \varphi\|^2_{H^r(\Omega)} + \|\boldsymbol{\nabla} \times (E - \varphi)\|^2_{H^r(\Omega)} \right)^{1/2} \tag{3.4.47}$$

and both sides represent a norm for the quotient space. Indeed, the right-hand side is the standard norm for a quotient space. Concerning the left-hand side, we need only to prove the definiteness, i.e., if the left-hand side vanishes for a function E, then E must be in Q^p. Since the polynomial space is finite-dimensional, the infimum on the left-hand side is attained for some specific $\varphi \in Q^p$. Both terms are nonnegative so they both must vanish. Vanishing of the second term implies[12] that $\boldsymbol{\nabla} \times (E - \varphi) \in \mathcal{P}^{r-1}$. Vanishing of the first term implies that $E - \varphi \in \mathcal{P}^{r-1}$. Consequently, $E - \varphi \in Q^p$ and, therefore, $E \in Q^p$ as well.

Now comes a delicate point. We claim that the quotient space equipped with both norms is complete. For the norm on the right-hand side, this is a standard result for Banach spaces. For the norm on the left-hand side, we need to show it. Let $E_n \in H^r(\mathrm{curl}, \Omega)/Q^p$ be a Cauchy sequence. Then E_n is a Cauchy sequence in $H^r(\Omega)/Q^p$ and also $\boldsymbol{\nabla} \times E_n$ is a Cauchy sequence in $H^r(\Omega)/\boldsymbol{\nabla} \times Q^p$. By Corollary 3.11, both spaces equipped with the alternative norm implied by the seminorms are complete and, therefore, both E_n and $\boldsymbol{\nabla} \times E_n$ converge to some limits, say, E, F. The touchy point is to show that $F = \boldsymbol{\nabla} \times E$ modulo a polynomial in Q^p. Consider any multi-index $\alpha, |\alpha| = r$. We have

$$(D^\alpha E_n, \boldsymbol{\nabla} \times \psi) = (D^\alpha \boldsymbol{\nabla} \times E_n, \psi) \quad \forall \psi \in \mathcal{D}(\Omega).$$

For a given $\psi \in \mathcal{D}(\Omega)$, both sides are continuous functionals on our quotient space. Passing to the limit, we obtain

$$(D^\alpha E, \boldsymbol{\nabla} \times \psi) = (D^\alpha F, \psi) \quad \forall \psi \in \mathcal{D}(\Omega).$$

Consequently,

$$D^\alpha(\boldsymbol{\nabla} \times E - F) = 0 \qquad \text{for every } |\alpha| = r,$$

which shows that $\boldsymbol{\nabla} \times E - F \in \mathcal{P}^{r-1} \subset \boldsymbol{\nabla} \times Q^p$. We are done.

Consequently, the identity map is continuous when the quotient space is equipped with those two norms. By the Banach Theorem, the inverse map (the identity itself) must be continuous as well. Thus the reverse inequality holds with some multiplicative constant C. Finally, we have trivially (set $\varphi = 0$),

$$\inf_{\varphi \in Q^p} \left(\|E - \varphi\|^2_{H^r(\Omega)} + \|\boldsymbol{\nabla} \times (E - \varphi)\|^2_{H^r(\Omega)} \right)^{1/2}$$
$$\leq C \left(|E|^2_{H^r(\Omega)} + |\boldsymbol{\nabla} \times E|^2_{H^r(\Omega)} \right)^{1/2}. \qquad \square \tag{3.4.48}$$

In the same way we prove an analogous result for the $H(\mathrm{div})$ spaces.

Theorem 3.14 (Bramble–Hilbert argument for $H^r(\mathrm{div})$-norm). *Let Ω be a domain in $\mathbb{R}^N$ and let V^p be a subspace of $H(\mathrm{div}, \Omega)$ such that*

$$\mathcal{P}^{p-1} \subset V^p \quad \text{and} \quad \mathcal{P}^{p-1} \subset \boldsymbol{\nabla} \cdot V^p \tag{3.4.49}$$

for some $p > 0$. Let $r > 0$ and let $p \geq r$. There exists a constant $C > 0$, dependent upon r, such that

$$\inf_{\phi \in V^p} \left(\|v - \phi\|^2_{H^r(\Omega)} + \|\boldsymbol{\nabla} \cdot (v - \phi)\|^2_{H^r(\Omega)} \right)^{1/2} \leq C \left(|v|^2_{H^r(\Omega)} + |\boldsymbol{\nabla} \cdot v|^2_{H^r(\Omega)} \right)^{1/2} \tag{3.4.50}$$

for every $v \in H^r(\mathrm{div}, \Omega)$, where

$$H^r(\mathrm{div}, \Omega) := \{v \in H^r(\Omega) : \boldsymbol{\nabla} \cdot v \in H^r(\Omega)\}. \tag{3.4.51}$$

---

[12]$|u|_{H^r(\Omega)} = 0 \Rightarrow u \in \mathcal{P}^{r-1}(\Omega).$

3.4.2 ▪ H^1, $H(\mathrm{curl})$, and $H(\mathrm{div})$ h-Interpolation Estimates

We discuss the 3D case only. Let

$$x = h\xi + b \tag{3.4.52}$$

be the simplest element map with $h = h_K$ being the element size.

The Piola transforms imply the following scalings for the H^1-, $H(\mathrm{curl})$-, $H(\mathrm{div})$-, and L^2-conforming elements:

$$u = \hat{u}\,, \qquad E = h^{-1}\hat{E}\,, \qquad v = h^{-2}\hat{v}\,, \qquad f = h^{-3}\hat{f}\,. \tag{3.4.53}$$

L^2-projection estimate. Let $f \in H^r(K)$, and let $f_p = Pf$ be the L^2-projection onto space Y_p such that $\mathcal{P}_{p-1} \subset Y_p$, $p \geq r$. We have

$$
\begin{aligned}
\|f - f_p\|^2 &= h^{-6}h^3\|\hat{f} - \hat{f}_p\|^2 && \left(\begin{array}{c}\text{Piola transform and}\\\text{change of coordinates}\end{array}\right)\\[4pt]
&= h^{-3}\inf_{\hat{\varphi}\in\hat{Y}_p}\|(I - \hat{P})(\hat{f} - \hat{\varphi})\|^2 && \text{(shape functions preservation property)}\\[4pt]
&\leq h^{-3}\|I - \hat{P}\|^2\inf_{\hat{\varphi}\in\hat{Y}_p}\|\hat{f} - \hat{\varphi}\|^2_{L^2(\hat{K})} && \text{(continuity of } L^2\text{-projection)}\\[4pt]
&\leq h^{-3}\inf_{\hat{\varphi}\in\hat{Y}_p}\|\hat{f} - \hat{\varphi}\|^2_{H^r(\hat{K})} && (\|I - \hat{P}\| = 1)\\[4pt]
&\lesssim h^{-3}|\hat{f}|^2_{H^r(\hat{K})} && \text{(Bramble–Hilbert Argument)}\\[4pt]
&= h^{-3}h^6 h^{-3}h^{2r}|f|^2_{H^r(K)} = h^{2r}|f|^2_{H^r(K)} && \text{(scalings)}\,.
\end{aligned}
\tag{3.4.54}
$$

$H(\mathrm{div})$-interpolation estimate. Let $v \in H^r(\mathrm{div}, K)$ be a given function, and let $v_p = \Pi^{\mathrm{div}}v \in V^p$ denote its FE interpolant. We assume that $\mathcal{P}_{p-1} \subset V^p$, $\mathcal{P}_{p-1} \subset \boldsymbol{\nabla}\cdot V^p, p \geq r$.

$$
\begin{aligned}
\|v - v_p\|^2 &= h^{-4}h^3\|\hat{v} - \hat{v}_p\|^2 && \left(\begin{array}{c}\text{scalings and change}\\\text{of variables}\end{array}\right)\\[6pt]
&= h^{-1}\|(I - \hat{\Pi}^{\mathrm{div}})\hat{v}\|^2 &&\\[6pt]
&= h^{-1}\inf_{\hat{\phi}\in\hat{V}^p}\|(I - \hat{\Pi}^{\mathrm{div}})(\hat{v} - \hat{\phi})\|^2 && \left(\begin{array}{c}\text{shape functions}\\\text{preservation property}\end{array}\right)\\[6pt]
&\lesssim h^{-1}\|I - \hat{\Pi}^{\mathrm{div}}\|^2_{\mathcal{L}(H^r(\mathrm{div},\hat{K}),L^2(\hat{K}))} &&\\[4pt]
&\quad\inf_{\hat{\phi}\in\hat{V}^p}\left(\|\hat{v} - \hat{\phi}\|^2_{H^r(\hat{K})} + \|\boldsymbol{\nabla}\cdot(\hat{v} - \hat{\phi})\|^2_{H^r(\hat{K})}\right) && \left(\begin{array}{c}\text{continuity of}\\\text{interpolation operator}\end{array}\right)\\[6pt]
&\lesssim h^{-1}\left(|\hat{v}|^2_{H^r(\hat{K})} + |\boldsymbol{\nabla}\cdot\hat{v}|^2_{H^r(\hat{K})}\right) && \left(\begin{array}{c}\text{Bramble–Hilbert}\\\text{Argument}\end{array}\right)\\[6pt]
&= h^{-1}\left(h^{2r+1}|v|^2_{H^r(K)} + h^{2r+3}|\boldsymbol{\nabla}\cdot v|^2_{H^r(K)}\right) && \text{(scalings)}\\[6pt]
&\leq h^{2r}|v|^2_{H^r(\mathrm{div},K)} && \left(\begin{array}{c}\text{definition of}\\\text{the seminorm}\end{array}\right)\,.
\end{aligned}
\tag{3.4.55}
$$

Notice that the higher power of h that we get in the second term is useless as the first term dominates.

The commuting diagram property implies now the estimate in the full $H(\text{div})$-norm. Indeed, for $f = \nabla \cdot v$, $\nabla \cdot \Pi^{\text{div}} v = Pf = f_p$, which implies that

$$\|\nabla \cdot (v - v_p)\|^2 = \|\nabla \cdot v - f_p\|^2 \leq Ch^{2r}|\nabla \cdot v|^2_{H^r(K)} \leq Ch^{2r}|v|^2_{H^r(\text{div},K)} . \quad (3.4.56)$$

Combining the two estimates above, we obtain

$$\|v - v_p\|^2_{H(\text{div},K)} \leq Ch^{2r}|v|^2_{H^r(\text{div},K)} . \quad (3.4.57)$$

$H(\text{curl})$-interpolation estimate. Let $E \in H^r(\text{curl}, K)$, and let $E_p = \Pi^{\text{curl}} E \in Q^p$ denote its FE interpolant. We assume that $\mathcal{P}^{p-1} \subset Q^p, \mathcal{P}^{p-1} \subset \nabla \times Q^p, p \geq r$, and proceed analogously to the $H(\text{div})$ case.

$$\|E - E_p\|^2 = h^{-2}h^3\|\hat{E} - \hat{E}_p\|^2 \qquad \left(\begin{array}{c} \text{scalings and} \\ \text{change of variables} \end{array}\right)$$

$$= h \inf_{\hat{\varphi} \in \hat{Q}^p} \|(I - \hat{\Pi}^{\text{curl}})(\hat{E} - \hat{\varphi})\|^2 \qquad \left(\begin{array}{c} \text{FE shape functions} \\ \text{preservation property} \end{array}\right)$$

$$\lesssim h \|I - \hat{\Pi}^{\text{curl}}\|^2_{\mathcal{L}(H^r(\text{curl},\hat{K}),L^2(\hat{K}))}$$

$$\inf_{\hat{\varphi} \in \hat{Q}^p} \left(\|\hat{E} - \hat{\varphi}\|^2_{H^r(\hat{K})} + \|\nabla \times (\hat{E} - \hat{\varphi})\|^2_{H^r(\hat{K})}\right) \qquad \left(\begin{array}{c} \text{continuity of} \\ \text{interpolation operator} \end{array}\right)$$

$$\lesssim h (|\hat{E}|^2_{H^r(\hat{K})} + |\hat{\nabla} \times \hat{E}|^2_{H^r(\hat{K})}) \qquad \left(\begin{array}{c} \text{Bramble–Hilbert} \\ \text{Argument} \end{array}\right)$$

$$= h (h^{2r-1}|E|^2_{H^r(K)} + h^{2r+1}|\nabla \times E|^2_{H^r(K)}) \qquad \text{(scalings)}$$

$$= h^{2r}|E|^2_{H^r(\text{curl},K)} \qquad \left(\begin{array}{c} \text{definition of} \\ \text{the seminorm} \end{array}\right) .$$

$$(3.4.58)$$

The commuting diagram property implies now the estimate in the full $H(\text{curl})$-norm. Indeed, for $v = \nabla \times E$, $\Pi^{\text{div}} v = v_p = \nabla \times E_p$, which implies that

$$\|\nabla \times (E - E_p)\|^2 = \|\nabla \times E - v_p\|^2 \leq Ch^{2r}|\nabla \times E|^2_{H^r(K)} \leq Ch^{2r}|E|^2_{H^r(\text{curl},K)} . \quad (3.4.59)$$

Combining the two estimates above, we obtain

$$\|E - E_p\|^2_{H(\text{curl},K)} \leq Ch^{2r}|E|^2_{H^r(\text{curl},K)} . \quad (3.4.60)$$

H^1-interpolation estimate. Let $u \in H^r(K)$, and let $u_p = \Pi^{\text{grad}} u \in W^p$ denote its FE interpolant. We assume that $\mathcal{P}^p \subset W^p, p+1 \geq r$. We have

$$\|u - u_p\|^2 = h^3\|\hat{u} - \hat{u}_p\|^2 \qquad \text{(scalings and change of variables)}$$

$$= h^3 \inf_{\hat{\varphi} \in \hat{W}^p} \|(I - \hat{\Pi}^{\text{grad}}))(\hat{u} - \hat{\varphi})\|^2 \qquad \text{(FE shape functions preservation property)}$$

$$\lesssim h^3 \|I - \hat{\Pi}^{\text{grad}}\|^2_{\mathcal{L}(H^r(\hat{K}),L^2(\hat{K}))}$$

$$\inf_{\hat{\varphi} \in \hat{W}^p} \|\hat{u} - \hat{\varphi}\|^2_{H^r(\hat{K})} \qquad \text{(continuity of interpolation operator)}$$

$$\lesssim h^3 |\hat{u}|^2_{H^r(\hat{K})} \qquad \text{(Bramble–Hilbert Argument)}$$

$$= h^{2r}|u|^2_{H^r(K)} \qquad \text{(scalings)} .$$

$$(3.4.61)$$

Assume now that $p \geq r$. Consequently, $p + 1 \geq r + 1$, and we can replace r with $r + 1$ to get

$$\|w - \Pi^{\mathrm{grad}} w\|_{L^2(K)} \lesssim h^{r+1} \|w\|_{H^{r+1}(K)}. \tag{3.4.62}$$

Moreover, applying the $H(\mathrm{curl})$ estimate to gradient ∇w, and using the commutativity argument, we get

$$\|\nabla w - \Pi^{\mathrm{curl}} \nabla w\| = \|\nabla(w - \Pi^{\mathrm{grad}} w)\| \lesssim h^r \|\nabla w\|_{H^r(K)} \leq h^r \|w\|_{H^{r+1}(K)}, \tag{3.4.63}$$

which yields the final estimate in the full norm,

$$\|w - \Pi^{\mathrm{grad}} w\|_{H^1(K)} \lesssim h^r \|w\|_{H^{r+1}(K)}. \tag{3.4.64}$$

Note that the L^2-interpolation error converges one order faster than the H^1 error. This is *not the case* for the $H(\mathrm{curl})$ and $H(\mathrm{div})$ estimates where the L^2-estimates and the corresponding energy estimates are of the same order.

Limited regularity case.   We explain the issue for the $H(\mathrm{curl})$ case only. The other cases are fully analogous. Two situations are possible:

- The interpolated function is (relative to p) regular, i.e., $p < r$. We use the estimate above with p in place of r to obtain

$$\|E - E_p\|_{H(\mathrm{curl},K)} \leq Ch^p \|E\|_{H^p(\mathrm{curl},K)} \leq Ch^p \|E\|_{H^r(\mathrm{curl},K)}. \tag{3.4.65}$$

  The rate of convergence is dictated by the polynomial order p.

- The interpolated function is less regular, $p > r$. We use the original estimate to obtain

$$\|E - E_p\|_{H(\mathrm{curl},K)} \leq Ch^r \|E\|_{H^r(\mathrm{curl},K)}. \tag{3.4.66}$$

  In this case, the rate of convergence is dictated by the regularity of the solution.

We usually combine the two estimates into one by writing

$$\|E - E_p\|_{H(\mathrm{curl},K)} \leq Ch^{\min\{p,r\}} \|E\|_{H^r(\mathrm{curl},K)}. \tag{3.4.67}$$

Remark 3.15. All the estimates above have been carried out for an integer r. An interpolation argument for Hilbert spaces can be used to generalize the results to real values of r.

Minimum regularity of interpolated functions.   What is the minimum value of r for which the standard interpolation operators are continuous on Sobolev spaces? It is sufficient to determine the minimum r for which the d.o.f. involved in the definition of the interpolation operators are well-defined. In H^1-interpolation, we use point values (e.g., at vertices or Lagrange nodes). The answer comes then from the Sobolev Embedding Theorem: $r > 1/2$ in 1D, $r > 1$ in 2D, and $r > 3/2$ in 3D. As $\nabla H^r \subset H^{r-1}(\mathrm{curl})$, the commuting diagram property implies that we must have $r > 0$ in 2D and $r > 1/2$ in 3D for computing the edge averages of E_t. This is indeed the case, and the estimate comes this time from trace theorems. In 2D, for $r > 0$, the trace of functions from $H^r(\mathrm{curl}, K)$ to an edge e lives in $H^{r-1/2}(e)$ and this is a sufficient regularity to compute the edge average. Indeed, the edge average of tangential component E_t can be viewed as action of E_t on the unity function,

$$\int_e E_t = \langle E_t, 1 \rangle,$$

and we need only to argue that the unity function lives in the dual of $H^{r-1/2}(e)$ for $r > 0$. This is indeed the case although the proof of this innocent statement requires a working knowledge of Sobolev spaces. In 3D we need to apply the Trace Theorem twice. For $r > 1/2$, trace E_t of a function E from $H^r(\text{curl}, K)$ to a face f lives in $H^{r-1/2}(\text{curl}, f)$. Applying the 2D result to the space on the face then finishes the reasoning. Finally, the Trace Theorem for $H(\text{div})$ spaces implies that that the face averages are well-defined for functions $v \in H^r(\text{div}, K)$ for $r > 0$. Indeed, the normal trace $v \cdot n$ to a face f then lives in $H^{r-1/2}(f)$, and this is again sufficient to interpret the face average of $v_n := v \cdot n$ as the action of v_n on the unity function.

The presented *PB interpolation increases the regularity assumptions.* For instance, the edge projections in 3D require the edge trace of a function u to be in $H^1(e)$. The Trace Theorem implies then that function u must come from $H^r(K)$ with $r > 2$. An interpolation argument in Sobolev spaces (see [57, Section 4.3]) shows that $u \in H^2(K)$ is sufficient. This still may be too demanding from the point of view of expected regularity of functions to be interpolated (exact solutions) and has led to a modification of the PB interpolation using fractional norms, compare [36] with [30, 26]. The version of the PB interpolation using projections in fractional norms requires the same regularity assumptions as the classical interpolation operators for the lowest order elements. It is more of an analysis tool than a practical algorithm to be implemented in a finite element code.

Estimates for general affine elements. Shape regularity assumptions. The presented interpolation error estimates generalize easily to the case of a general affine isomorphism,

$$x_K : \hat{K} \ni \xi \to x = A\xi + x_0 \,.$$

Here A is a nonsingular matrix, $\det A \neq 0$, and x_0 is a point. Obviously, both may depend upon the element K. Note that the inverse of an affine isomorphism is an affine isomorphism as well. Typically, we request $\det A > 0$.

In place of simple scalings, we need now more careful estimates for the Piola maps. We have

$$j = \det A, \quad j^{-1} = \det A^{-1} \,,$$

$$\|E\| \leq \|J^{-T}\| \,\|\hat{E}\| = \|A^{-1}\| \,\|\hat{E}\| \,,$$

$$\|H\| \leq |\det A^{-1}| \,\|A\| \,\|\hat{H}\| \,,$$

$$|f| \leq |\det A^{-1}| \,|\hat{f}| \,,$$

where all norms of vectors and matrices are Euclidean norms. We may estimate them in terms of geometrical quantities. For instance,

$$\|A\| \leq \frac{h}{\hat{\rho}} \,, \tag{3.4.68}$$

where $h = h_K$ is the element size defined as

$$h_K := \sup_{x,y \in K} \|x - y\| \,,$$

and $\hat{\rho}$ is the diameter of the largest sphere contained in the corresponding master element $\hat{K}$; compare Exercise 3.4.3.

In the same way we can estimate higher order Sobolev seminorms. Start with the transformation rule for the first order differential,

$$d_\xi \hat{u}(\hat{e}) = d_x u(A\hat{e}) \,.$$

Analogously, for a differential of order r,

$$d_\xi^r \hat{u}(\hat{e}_1, \ldots, \hat{e}_r) = d_x^r u(A\hat{e}_1, \ldots, A\hat{e}_r). \tag{3.4.69}$$

Consequently,

$$\|d_\xi^r \hat{u}\| := \sup_{\hat{e}_1, \ldots, \hat{e}_r} \frac{d_\xi^r \hat{u}(\hat{e}_1, \ldots, \hat{e}_r)}{\|\hat{e}_1\|, \ldots, \|\hat{e}_r\|}$$

$$= \sup_{\hat{e}_1, \ldots, \hat{e}_r} \frac{d_x^r u(A\hat{e}_1, \ldots, A\hat{e}_r)}{\|A\hat{e}_1\|, \ldots, \|A\hat{e}_r\|} \frac{\|A\hat{e}_1\|, \ldots, \|A\hat{e}_r\|}{\|\hat{e}_1\|, \ldots, \|\hat{e}_r\|}$$

$$\leq \|d_x^r u\| \, \|A\|^r.$$

In conclusion, if we can bound $\|A\|$, $\|A^{-1}\|$, $|j^{-1}|$ *uniformly for all elements in the mesh*, all the discussed interpolation error estimates hold as well at the expense of introducing additional constants reflecting shape regularity; compare Exercise 3.4.4.

Note that formula (3.4.69) does not hold for a nonconstant Jacobian. In the case of a general element map, the rth derivative in ξ will depend upon not only rth derivative in x but also all derivatives of lower order $1, \ldots, r-1$. Consequently, the scaling argument fails and the estimates *do not* generalize to nonaffine elements.

Discretization in the parametric domain.   *This is very important.* The element maps need not be random (as it happens, for instance, in the case of unstructured mesh generators). In the case of CAD defined geometries, they come from a predefined global geometry map in a reference domain (see Fig. 3.5), where element map x_K is the composition of an affine reference map $\eta = \eta(\xi)$ and the CAD parametrization $x = x(\eta)$. The entire boundary-value problem can then be redefined in the reference domain. The geometry maps contribute then to the redefined material data, and the original problem is effectively solved in the reference domain where all elements are shape regular affine elements. The CAD parametrizations can be used directly (exact geometry elements) or they can be approximated (interpolated) with polynomials, usually coming from the H^1-space of element shape functions W^p (isoparametric element).

Analyzing the convergence in the reference rather than the physical domain comes with a price, though. The Sobolev norms of the solution present in the interpolation error estimates involve now the *composition of CAD maps with the actual solution defined in the physical domain*. The regularity of the CAD maps affects the regularity of the compositions. If the CAD maps are poor, with low regularity, the overall regularity of the compositions will also be low, which will affect the convergence rates.

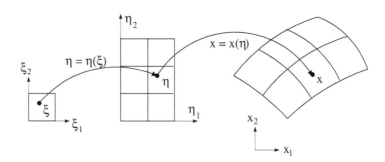

Figure 3.5: Reference geometry map.

3.4.3 ▪ hp-Interpolation Estimates

If the interpolation operator preserves FE shape functions, and we have at our disposal p-interpolation estimates on the master element, we can immediately use the discussed scaling arguments to obtain the corresponding hp-interpolation error estimates. For instance, for the $H(\text{curl})$ case, we have [30, 26]

$$\|\hat{E} - \hat{\Pi}^{\text{curl}}\hat{E}\|_{H(\text{curl},\hat{K})} \leq C(r) \ln p\, p^{-r} \|\hat{E}\|_{H^r(\text{curl},\hat{K})} \,. \qquad (3.4.70)$$

Instead of using the continuity of the interpolation operator, we can use now the p-estimate,

$$
\begin{aligned}
&h^1 \|\hat{E} - \hat{\Pi}^{\text{curl}}\hat{E}\|^2_{H(\text{curl},\hat{K})} \\
&= h^1 \inf_{\hat{\psi} \in Q^p} \|(\hat{E} - \hat{\psi}) - \hat{\Pi}^{\text{curl}}(\hat{E} - \hat{\psi})\|^2_{H(\text{curl},\hat{K})} \quad \text{(shape functions preservation)} \\
&\lesssim h^1 \ln^2 p\, p\, p^{-2r} \inf_{\hat{\psi} \in \hat{Q}_p} \|\hat{E} - \hat{\psi}\|_{H^r(\text{curl},\hat{K})} \qquad \text{(p-interpolation error estimate)} ,
\end{aligned}
\qquad (3.4.71)
$$

with the rest of the argument remaining identical as in the h-case. The ultimate estimate reads as follows:

$$\|u - \Pi^{\text{curl}}E\|_{H(\text{curl},K)} \leq C(r) \ln p\, \frac{h^{\min\{p,r\}}}{p^r} \|E\|_{H^r(\text{curl},K)} \,. \qquad (3.4.72)$$

In a similar way, we obtain the remaining hp-interpolation error estimates,

$$
\begin{aligned}
\|u - \Pi^{\text{grad}}u\|_{H^1(K)} &\leq C(r) \ln^2 p\, \frac{h^{\min\{p,r\}}}{p^r} \|u\|_{H^{r+1}(K)} , \\
\|v - \Pi^{\text{div}}v\|_{H(\text{div},K)} &\leq C(r) \ln p\, \frac{h^{\min\{p,r\}}}{p^r} \|v\|_{H^r(\text{div},K)} , \\
\|f - Pf\|_{L^2(K)} &\leq C(r) \frac{h^{\min\{p,r\}}}{p^r} \|f\|_{H^r(K)} \,.
\end{aligned}
\qquad (3.4.73)
$$

Exercises

3.4.1. Prove Lemma 3.8. *Hint:* Revisit the proof of Lemma 2.3. (2 points)

3.4.2. Norm of a matrix induced by Euclidean norm. Let $A \in L(\mathbb{R}^n, \mathbb{R}^n)$ be a linear map. Let $\|\cdot\|$ be the standard l^2 (Euclidean) norm in $\mathbb{R}^n$. The corresponding induced norm for map A is defined as

$$\|A\| := \sup_{x \neq 0} \frac{\|Ax\|}{\|x\|} \,.$$

   (i) Demonstrate that the norm of A equals the maximum characteristic value of A:

$$\|A\| = \max_{i=1,\ldots,n} \lambda_i,$$

   where $\lambda_i \geq 0$, λ_i^2 are eigenvalues of AA^T or, equivalently, $A^T A$.

   (ii) Extend the formula to the complex case.

   (5 points)

3.4.3. Estimate of the Euclidean norm of a linear map (Jacobian of an affine map). Prove estimate (3.4.68). (3 points)

3.4.4. Interpolation error estimates for an affine element. Rederive all four interpolation error estimates for a general affine element. Use geometrical estimate (3.4.68) for Jacobians. (10 points)

3.4.5. Fractional Sobolev spaces. Consider the infinite L-shaped domain shown in Fig. 3.6.

    (i) Switch to polar coordinates and use separation of variables to determine a family of solutions to the Laplace equation with homogeneous BC $u = 0$ on the reentrant edges.

    (ii) Determine the most singular solution that belongs to the energy space H^1_{loc} (it is in H^1 in any compact neighborhood of the reentrant corner). It should be in the form of

$$u(r, \theta) = r^\alpha f(\theta) \,,$$

    where $f(\theta)$ is a smooth function.

    (iii) Determine values of exponent α for which the function above lives in H^1_{loc} or H^2_{loc}. Guess the fractional Sobolev space in which the actual solution lives.

This "guessing" procedure may be made very precise using the interpolation theory for Sobolev spaces. Solution to a corresponding Laplace problem in a bounded domain containing the reentrant corner (with the same BC along the reentrant edges but arbitrary BC on the remaining part of the boundary) will have the same singularity at the corner. The solution you have developed is commonly used as a manufactured solution for a bounded domain to verify the expected convergence rates.
(5 points)

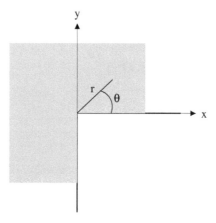

Figure 3.6: Infinite L-shaped domain.

3.5 ▪ Aubin–Nitsche Argument

Consider a variational problem satisfying the assumptions of the Lax–Milgram Theorem, and its Bubnov–Galerkin discretization.

$$\begin{cases} u \in U, \\ b(u, v) = l(v), \quad v \in U, \end{cases} \qquad \rightarrow \qquad \begin{cases} u_h \in U_h \subset U, \\ b(u_h, v_h) = l(v_h), \quad v_h \in U_h. \end{cases}$$

Let M and α denote the continuity and coercivity constants for the bilinear form. Cea's Lemma argument establishes convergence in the energy norm,

$$\|u - u_h\|_U \leq \frac{M}{\alpha} \inf_{w_h \in U_h} \|u - w_h\|_U \,,$$

with the rate of the *best approximation error* measured in the energy norm. For problems with the H^1 energy norm setting, this does not imply an optimal convergence rate in the *weaker* L^2-norm. The optimal rate of convergence in the L^2-norm can be established using a duality argument known as the *Aubin–Nitsche trick*. Consider the dual problem

$$\begin{cases} v_g \in U \,, \\ b(w, v_g) = (w, \underbrace{u - u_h}_{=:g}) \quad \forall w \in U \,, \end{cases}$$

where $(\cdot, \cdot)$ is the L^2-inner product. Assume that the dual problem is well-posed and admits a stability estimate in a Sobolev norm *stronger* than the energy norm H^1:

$$\|v_g\|_{H^{1+s}(\Omega)} \leq C\|g\|, \quad s > 0 \,. \tag{3.5.74}$$

A stability estimate of this type is a consequence of a *regularity result* and the Banach Theorem argument. The dual variational problem is still set up in the same energy space U and, a priori, we control only solution u in the energy $\|\cdot\|_U$-norm. Due to the more regular load $g \in L^2(\Omega) \subset U'$ though, for a sufficiently regular domain and material data, the solution is typically more regular than its energy space setting. Consider the strong form of the map corresponding to the dual problem. As the map

$$B' : H^{1+s}(\Omega) \ni v \to B'v \in L^2(\Omega)$$

is well-defined and, by the postulated regularity result, surjective, the Banach Theorem implies that its inverse must be continuous as well, i.e., we arrive at the stability estimate (3.5.74). For a standard elliptic problem and smooth or convex Lipschitz domain Ω, $s = 1$. In the case of a Lipschitz domain with corners and edges, $s < 1$ but always positive. We have then

$$\begin{aligned} \|u - u_h\|^2 &= (u - u_h, u - u_h) \\ &= b(u - u_h, v_g) && \text{(definition of the dual problems)} \\ &= b(u - u_h, v_g - v_h) && \text{(Galerkin orthogonality)} \\ &\leq M\|u - u_h\|_U \|v_g - v_h\|_U && \text{(continuity)} \\ &\leq CM\|u - u_h\|_U \, h^s\|v_g\|_{H^{1+s}} && \text{(best approximation error estimate for } v_g) \\ &\leq CMh^s\|u - u_h\|_U \|u - u_h\| \,. \end{aligned}$$

Dividing both sides by $\|u - u_h\|$, we obtain

$$\|u - u_h\| \leq CMh^s\|u - u_h\|_U \,.$$

Thus, if the solution u_h converges to u with a specific rate h^r in the energy norm, it will converge also to u in the L^2-norm with *a higher rate* h^{r+s}. The gain s depends upon the stability properties of the *continuous* dual problem. For standard second order elliptic problems and smooth or convex domains, $s = 1$, i.e., the actual L^2-error converges with the same rate as the best approximation error.

3.5.1 ▪ Generalizations

The duality argument can be extended to more complicated projections. We will discuss now a few examples stemming from the study of two-grid methods for $H(\mathrm{div})$-projections and linear acoustics [3]. For simplicity of the argument we assume convexity of domain Ω.

Weighted H^1-norm. Consider the norm on $H^1(\Omega)$,

$$\|u\|_E^2 := \|\nabla u\|^2 + \alpha^2 \|u\|^2 \,,$$

parametrized with $\alpha \geq 0$. For $\alpha = 0$ we are back to the Laplace equation. For large α, we arrive at a reaction-dominated diffusion problem. The goal is to repeat our duality argument to obtain an L^2-error estimate showing explicit dependence upon parameter α. The dual problem coincides with the original problem,

$$\begin{cases} v_y \in H_0^1(\Omega)\,, \\ (\nabla \delta u, \nabla v_g) + \alpha^2(\delta u, v_g) = (\delta u, g)\,, \quad \delta u \in H_0^1(\Omega)\,, \end{cases}$$

with the corresponding strong form,

$$-\Delta v_g + \alpha^2 v_g = g \,.$$

Substituting $\delta u = v_g$, we obtain the standard stability estimate,

$$\|v_g\|_E^2 \leq \|g\|\,\|v_g\| \leq C_P \|g\|\,\|\nabla v_g\| \leq C_P \|g\|\,\|v_g\|_E \,,$$

where C_P is the Poincaré constant. This implies

$$\alpha \|v_g\| \leq C_P \|g\| \,.$$

Notice that, for $\alpha \geq 1$, we can replace the Poincaré constant with one (explain why). The strong form of the dual problem now implies

$$\|\Delta v_g\| \leq \|g\| + \alpha^2 \|v_g\| \lesssim (1 + \alpha)\|g\| \,.$$

We can use a standard elliptic regularity argument to conclude that

$$\|v_g\|_{H^2(\Omega)} \lesssim (1 + \alpha)\|g\| \,.$$

The Aubin–Nitsche duality argument implies now that

$$\|u - u_h\|^2 \leq \|u - u_h\|_E \, \|v_g - \Pi_h v_g\|_E \,,$$

where Π_h is an H^1-interpolation operator. With the solution of the dual problem in $H^2(\Omega)$, we estimate the interpolation error as follows:

$$\|v_g - \Pi_h v_g\|_E \leq \|\nabla(v_g - \Pi_h v_g)\| + \alpha\|v_g - \Pi_h v_g\| \lesssim h(1 + \alpha h)\|v_g\|_{H^2(\Omega)} \,.$$

Combining all arguments together, we obtain the final estimate of the L^2-error:

$$\|u - u_h\| \lesssim \|u - u_h\|_E \,(1 + \alpha h)(1 + \alpha)h \,. \tag{3.5.75}$$

In particular, for $u \in H_0^1(\Omega)$, we obtain

$$\|u - u_h\| \lesssim (1 + \alpha h)(1 + \alpha)h\|u\|_E \,.$$

Remark 3.16. For $\alpha = 0$, we recover the standard estimate. For $\alpha > 0$, asymptotically in h, i.e., for $\alpha h \leq 1$, we see a linear dependence of the stability constant upon α. Note that the duality argument makes sense only in the asymptotic regime. For $\alpha h > 1$, the simple energy stability argument gives a better estimate,

$$\|u - u_h\| \leq \frac{1}{\alpha}\|u - u_h\|_E \leq \frac{1}{\alpha}\|u\|_E = \frac{1}{\alpha h}h\|u\|_E \leq h\|u\|_E \,.$$

The bigger the αh, the smaller the stability constant.

Weighted $H(\mathrm{div})$-norm. Consider the energy norm on $H(\mathrm{div}, \Omega)$,

$$\|u\|_E^2 := \|\operatorname{div} u\|^2 + \alpha^2\|u\|^2 \,,$$

where $\alpha \geq 0$. We shall attempt to generalize the duality argument to the weighted $H(\mathrm{div})$-projection:

$$\begin{cases} u_h \in V_h \subset H_0(\mathrm{div}, \Omega) \,, \\ (\operatorname{div}(u_h - u), \operatorname{div} v_h) + \alpha^2(u_h - u, v_h) = 0 \,, \quad v_h \in V_h \subset H_0(\mathrm{div}, \Omega) \,. \end{cases}$$

Employing $v_h = \boldsymbol{\nabla} \times \phi_h$, $\phi_h \in Q_h \subset H_0(\mathrm{curl}, \Omega)$, we learn that

$$(u_h - u, \boldsymbol{\nabla} \times \phi_h) = 0 \,, \quad \phi_h \in Q_h \subset H_0(\mathrm{curl}, \Omega) \,,$$

where Q_h denotes the $H_0(\mathrm{curl}, \Omega)$ member of the discrete exact sequence. It makes thus sense to assume that, for $\alpha = 0$, we have a *constrained projection problem*:

$$\begin{cases} \|\operatorname{div}(u_h - u)\| \to \min_{u_h \in V_h} \,, \\ (u_h - u, \boldsymbol{\nabla} \times \phi_h) = 0 \,, \quad \phi_h \in Q_h \,. \end{cases}$$

This is exactly the projection operator P_h^{div}, member of a family of commuting projection operators introduced in [49]. Note that the operator coincides exactly with the construction from [3].

Clearly, we do not expect the higher rate of convergence for any function $u \in H(\mathrm{div}, \Omega)$ at least for two reasons: (a) the projection involves only a combination of derivatives (the divergence), (b) the $H(\mathrm{div})$-conforming space V_h does not include complete polynomials of order p, just some of them. For $u = \boldsymbol{\nabla} \times \psi$, $\psi \in H_0(\mathrm{curl}, \Omega)$, we have $\operatorname{div} u_h = \operatorname{div} u = 0$. Consequently, projection u_h is itself a curl, and the constrained projection problem reduces to the projection problem in $H_0(\mathrm{curl}, \Omega)$,

$$(\boldsymbol{\nabla} \times (\psi_h - \psi), \boldsymbol{\nabla} \times \phi_h) = 0 \,, \quad \phi_h \in Q_h \quad \Leftrightarrow \quad \|\boldsymbol{\nabla} \times (\psi_h - \psi)\| \to \min_{\psi_h \in Q_h} \,.$$

The minimizer ψ_h is not unique but $\boldsymbol{\nabla} \times \psi_h$ is; see [49] for the construction of commuting projection operators. Consequently,

$$\|u_h - u\| = \|\boldsymbol{\nabla} \times (\psi_h - \psi)\| \leq \|\boldsymbol{\nabla} \times \psi\| = \|u\|_{H(\mathrm{div}, \Omega)}$$

with the estimate above being sharp (orthogonal projection). This means that for $u = \boldsymbol{\nabla} \times \psi$, we cannot expect a better convergence rate in the L^2-norm.

Lemma 3.17 (Helmholtz Decomposition). *Let Ω be a simply connected domain. We have the corresponding orthogonal decomposition,*

$$H_0(\mathrm{div}, \Omega) = \boldsymbol{\nabla} \times H_0(\mathrm{curl}, \Omega) \overset{\perp}{\oplus} \boldsymbol{\nabla} H^1(\Omega) \,.$$

The two subspaces are orthogonal in both the L^2 and the $H(\mathrm{div})$ sense.

Proof. Let $p \in H^1(\Omega)$ be the solution of the problem

$$(\nabla p, \nabla q) = (u, \nabla q), \quad q \in H^1(\Omega).$$

Equivalently, $\Delta p = \operatorname{div} u$ with homogeneous Neumann BC. The problem is well-defined, and potential p is unique up to an additive constant. Then $\operatorname{div}(u - \nabla p) = 0$ and, by the exact sequence property, there exists a (nonunique) vector potential $\psi \in H_0(\operatorname{curl}, \Omega)$ such that $u - \nabla p = \nabla \times \psi$. Orthogonality follows from integration by parts, and the direct sum decomposition is a consequence of the orthogonality. □

The Helmholtz decomposition result motivates us to restrict our considerations to $u = \nabla p$, $p \in H^1(\Omega)$, $\operatorname{div} u = \Delta p \in L^2(\Omega)$, $u_n = \partial p / \partial n = 0$ on Γ. Let $u_h =: P_h^{\operatorname{div}} u$ be the orthogonal projection of u in the energy norm, with the corresponding Helmholtz decomposition,

$$u_h = \nabla \times \psi^h + \nabla p^h. \tag{3.5.76}$$

Notice the use of upper indices[13] for the potentials that are *not* in the discrete spaces. Consequently,

$$u_h - u = \nabla \times \psi^h + \nabla(p^h - p).$$

We shall estimate the two terms separately. The second term is estimated using the duality arguments. Let $g := \nabla(p^h - p)$. Define the dual problem,

$$\begin{cases} v_g \in H_0(\operatorname{div}, \Omega), \\ (\operatorname{div} \delta v, \operatorname{div} v_g) + \alpha^2 (\delta v, v_g) = (\delta v, g), & \delta v \in H_0(\operatorname{div}, \Omega), \end{cases}$$

or, in the strong form,

$$-\nabla \operatorname{div} v_g + \alpha^2 v_g = g.$$

Substituting $\delta v := v_g$ and using the Friedrichs inequality, we obtain

$$\| \operatorname{div} v_g \|^2 + \alpha^2 \|v_g\|^2 = (g, v_g) \leq \|g\| \, \|v_g\| \leq C_F \|g\| \, \| \operatorname{div} v_g \|$$

$$\leq C_f \|g\| \left(\| \operatorname{div} v_g \|^2 + \alpha^2 \|v_g\|^2 \right)^{\frac{1}{2}},$$

where C_F is the Friedrichs constant. This implies

$$\| \operatorname{div} v_g \| \leq C_F \|g\|$$

and

$$\alpha \|v_g\| \leq C_F \|g\|.$$

The strong form of the dual problem then implies

$$\|\nabla \operatorname{div} v_g\| \lesssim (1 + \alpha) \|g\|.$$

Testing with $\delta v = \nabla \times \phi$, $\phi \in H_0(\operatorname{curl}, \Omega)$, we learn that $v_g = \nabla \psi$, $\psi \in H^1(\Omega)$ with $\partial \psi / \partial n = 0$ on Γ. The inequality above ensures that

$$\|\Delta \psi\|_{H^1(\Omega)} \lesssim (1 + \alpha) \|g\|.$$

_____

[13] Kikuchi's notation.

If we assume additionally that domain Ω is $C^{1,1}$-regular, we can conclude that

$$\|\psi\|_{H^3(\Omega)} \lesssim (1+\alpha)\|g\|$$

and, therefore, $v_g \in H^2(\Omega) \cap H^1(\mathrm{div}, \Omega)$ with the norm controlled by $(1+\alpha)\|g\|$. The high regularity of the solution of the dual problem leads to the interpolation error estimate analogous to the elliptic case,

$$\begin{aligned}
\|v_g - \Pi_h^{\mathrm{div}} v_g\|_E^2 &= \|\mathrm{div}(v_g - \Pi_h^{\mathrm{div}} v_g)\|^2 + \alpha^2 \|v_g - \Pi_h^{\mathrm{div}} v_g\|^2 \\
&\lesssim h^2 \|v_g\|_{H^1(\mathrm{div},\Omega)}^2 + \alpha^2 h^4 \|v_g\|_{H^2(\Omega)}^2 \\
&\lesssim \left(h(1+\alpha h)(1+\alpha)\|g\| \right)^2 .
\end{aligned}$$

We can now use the duality argument:

$$\begin{aligned}
\|g\|^2 &= (u - u_h, g) = (\mathrm{div}(u - u_h), \mathrm{div}\, v_g) + \alpha^2 (u - u_h, v_g) \\
&\qquad\qquad\qquad\qquad\qquad\qquad \text{(definition of dual problem)} \\
&= (\mathrm{div}(u - u_h), \mathrm{div}\, v_g - \Pi_h^{\mathrm{div}} v_g) + \alpha^2 (u - u_h, v_g - \Pi_h^{\mathrm{div}} v_g) \\
&\qquad\qquad\qquad\qquad\qquad\qquad \text{(Galerkin orthogonality)} \\
&\leq \|u - u_h\|_E \, \|v_g - \Pi_h^{\mathrm{div}} v_g\|_E \\
&\qquad\qquad\qquad\qquad\qquad\qquad \text{(Cauchy–Schwarz)} \\
&\lesssim \|u - u_h\|_E \, (1+\alpha h)(1+\alpha)h\|g\| \\
&\qquad\qquad\qquad\qquad\qquad\qquad \text{(interpolation error estimate)} .
\end{aligned}$$

This leads to the final estimate of $\|g\|$ of the same form as for the elliptic case,

$$\|g\| \lesssim (1+\alpha h)(1+\alpha)h \, \|u - u_h\|_E .$$

In particular, for $u \in H(\mathrm{div}, \Omega)$ only, we get

$$\|g\| \lesssim (1+\alpha h)(1+\alpha)h \, \|u\|_E . \tag{3.5.77}$$

Remark 3.18. The estimate shows that, for a more regular domain and $\alpha h \leq 1$, the L^2-error depends linearly on α. If we are not concerned with the dependence upon α, but only in h, we can estimate the L^2-part of the energy norm of the interpolation error with the first power of h only which does not require the higher regularity of v_g.

We proceed now with the estimate of the remaining term $\|\nabla \times \psi^h\|$. Multiplying (3.5.76) with $\nabla \times \phi_h$, $\phi_h \in Q_h$, and using the fact that both u_h and ∇p^h are discrete curl-free, we conclude that so is $\nabla \times \psi^h$, i.e.,

$$(\nabla \times \psi^h, \nabla \times \phi_h) = 0, \quad \phi_h \in Q_h . \tag{3.5.78}$$

At the same time, interpolating both sides of (3.5.76) and utilizing the commuting property of the interpolation operators, we obtain

$$u_h = \Pi_h^{\mathrm{div}} u_h = \Pi_h^{\mathrm{div}}(\nabla \times \psi^h) + \Pi_h^{\mathrm{div}}(\nabla p^h) = \nabla \times (\Pi_h^{\mathrm{curl}} \psi^h) + \Pi_h^{\mathrm{div}}(\nabla p^h) .$$

Subtracting the result above from (3.5.76), we learn that

$$\|\boldsymbol{\nabla} \times (\psi^h - \Pi_h^{\mathrm{curl}}\psi^h)\| = \|\boldsymbol{\nabla} p^h - \Pi_h^{\mathrm{div}}(\boldsymbol{\nabla} p^h)\| . \tag{3.5.79}$$

This leads to the final estimate,

$$
\begin{aligned}
\|\boldsymbol{\nabla} \times \psi^h\| &\leq \left(\|\boldsymbol{\nabla} \times \psi^h\|^2 + \|\boldsymbol{\nabla} \times \Pi_h^{\mathrm{curl}}\psi^h\|^2 \right)^{1/2} \\
&= \|\boldsymbol{\nabla} \times (\psi^h - \Pi_h^{\mathrm{curl}}\psi^h)\| && \text{(orthogonality (3.5.78))} \\
&= \|\boldsymbol{\nabla} p^h - \Pi_h^{\mathrm{div}}(\boldsymbol{\nabla} p^h)\| && \text{(3.5.79)} \\
&\lesssim h\|p^h\|_{H^2(\Omega)} && \text{(interpolation error estimate)} \\
&\lesssim h\|\Delta p^h\| && \text{(elliptic regularity)} \\
&= h\|\operatorname{div} u_h\| \leq h\|u_h\|_E \leq h\|u\|_E && \text{(stability of energy projection)} .
\end{aligned}
$$

Note that there is no dependence upon α. On the negative side, we have managed to prove that this term converges to zero but, contrary to the other term, we have not managed to bound it with the product of the energy norm of the error and element size h. Finally, note that Remark 3.16 remains valid in this case, as well.

Linear acoustics. The arguments used for the weighted H^1- and $H(\operatorname{div})$-projections can be recycled to obtain a similar bound for the L^2-error of the energy projection in the acoustics graph norm,

$$\mathsf{u} := (u, p) \in H_0(\operatorname{div}, \Omega) \times H_0^1(\Omega) ,$$

$$\|\mathsf{u}\|_E^2 := \omega^2 \|\mathsf{u}\|^2 + \|A\mathsf{u}\|^2 ,$$

$$A\mathsf{u} := (i\omega u + \boldsymbol{\nabla} p, i\omega p + \operatorname{div} u) ,$$

where $\omega > 0$ is the frequency. Under the same assumptions as for the first two projections, we can show that, for $\mathsf{u} = (\boldsymbol{\nabla} q, p) \in H_0(\operatorname{div}, \Omega) \times H_0^1(\Omega)$, we have the estimate

$$\|\mathsf{u} - \mathsf{u}_h\| \lesssim (1 + \omega h)\omega h\|\mathsf{u}\|_E .$$

Exercises

3.5.1. Duality argument for the L^2-projection. Let $u \in L^2(\Omega)$ and let $u_h \in U_h$ be the L^2-projection of u onto a typical FE space U_h admitting the best approximation error estimate,

$$\inf_{v_h \in U_h} \|v - v_h\| \leq Ch^s \|v\|_{H^s(\Omega)}, \qquad v \in H^s(\Omega) .$$

Use the duality argument to show the improved rate of convergence in the dual norm,

$$\|u - u_h\|_{(H^s(\Omega))'} \leq Ch^s \|u - u_h\|$$

for any $s > 0$. Note that the result does not require any regularity assumptions on domain Ω besides those used to establish the best approximation error in the H^s-norm above. Finally, show that the dual norm is stronger than the negative norm $\|u - u_h\|_{H^{-s}(\Omega)}$. Consult [27, proof of Theorem 3.1.1] if necessary. (10 points)

3.6 ▪ Clément Interpolation

Both classical and PB interpolation are classified as *local interpolation* techniques. The element interpolant depends only upon the interpolated function and its derivatives within the element. This is the good news. The not so good news is that those interpolants are defined only for sufficiently regular functions forming a proper subspace of the energy space. In many applications, including construction of Fortin operators (see Section 4.3) and a posteriori error estimation [15], we are in need of interpolation operators defined *on the whole energy space*. The H^1-conforming Clément interpolation [19, 15] presented in this section falls into this category. We pay a price, though. The interpolation is no longer local—the element interpolant depends upon the function values evaluated outside of the element.

Let $\Omega \subset \mathbb{R}^N$ be a polygonal (polyhedral) domain partitioned into affine simplicial elements satisfying the usual mesh regularity assumptions.

Consider standard Lagrange elements of order p. For a Lagrangian node a_i, let S_i denote the support of the corresponding basis function e_i, a patch of elements sharing node a_i. Let $u \in L^2(\Omega)$, and let $u_i^p \in \mathcal{P}^p(S_i)$ be the L^2-projection of function u onto polynomials of order p on the element patch S_i,

$$\begin{cases} u_i^p \in \mathcal{P}^p(S_i)\,, \\ \displaystyle\int_{S_i} (u - u_i^p)\,\phi = 0 \quad \forall \phi \in \mathcal{P}^p(S_i)\,. \end{cases}$$

The Clément interpolation is defined similarly to the Lagrange interpolation except that pointwise values $u(a_i)$ are replaced with values of the corresponding L^2-projections $u_i^p(a_i)$,

$$\Pi_h u := \sum_i u_i^p(a_i)\phi_i\,,$$

where ϕ_i is the Lagrangian shape function corresponding to node a_i. Notice that the operator is only *semilocal*. For an element K, $\Pi_h u$ depends upon values of u in the patch of all elements adjacent to K.

Theorem 3.19 (Clément, 1975). *Let $u \in H^r(\Omega)$, $r \le p+1$. The following interpolation error estimates hold:*

$$|u - \Pi_h u|_{H^k(\Omega)} \le Ch^{r-k}|u|_{H^r(\Omega)}, \qquad k = 0, 1, \ldots, r\,, \tag{3.6.80}$$

where constant C is independent of u.

Note that, due to the nonlocality of Clément interpolation, estimate (3.6.82) is formulated globally. Following [19], we will prove the theorem for the 2D case. The 3D case is fully analogous.

Lemma 3.20. *Let S be a patch of triangular elements sharing a node. Let $u \in H^r(S)$, $r \le p+1$, and $u_p \in \mathcal{P}^p(S)$ be the L^2-projection of function u onto polynomials $\mathcal{P}^p(S)$. Equivalently,*

$$(u - u_p, \phi)_S = 0, \quad \phi \in \mathcal{P}^p(S)\,,$$

where $(\cdot, \cdot)_S$ is the $L^2(S)$ inner product. There exists then a constant C, independent of u, u_p, and patch S such that

$$|u - u_p|_{H^k(S)} \le C\,d(S)^{r-k}\,|u|_{H^r(S)}, \qquad 0 \le k \le r\,, \tag{3.6.81}$$

where $d(S)$ is the diameter of patch S.

A few comments first. In the case of a Lagrangian node interior to an element, the patch reduces to the single element. As the L^2-projection error is bounded by the interpolation error, the L^2-estimate follows from the standard interpolation theory. The estimates in higher order seminorms are already new. We learned from the Aubin–Nitsche theory that projection in H^1-seminorm yields optimal convergence rate in the L^2-norm. The result above says that the converse is true as well; given the regularity, the L^2-projection yields optimal h-convergence rates in higher order seminorms as well. In the case of a multielement patch, the additional nontriviality of the result is the fact that constant C is patch-independent. More precisely, given the mesh regularity assumptions, constants corresponding to different patches admit a common upper bound. In the following proofs, C will stand for a generic constant that depends upon the minimum angle of an element, number of elements in the patch, etc.

Proof. The arguments are rather technical. We will prove the case $d(S) = 1$. The general case follows then from the case $d(S) = 1$ and the usual scaling argument. Let $\hat{S}$ be a fixed *master patch* consisting of elements $\hat{K}$. In the case of an edge node, the patch consists of just two elements and we may select $\hat{S}$ shown in Fig. 3.7(a). In the case of a vertex patch, the patch will look different for a node located on boundary Γ (see, e.g., Fig. 3.7(a)) or for a vertex node from the interior of the domain (see, e.g., Fig. 3.7(b)). We need to select separate patches for $n = 3, 4, \ldots$ elements. With the mesh shape regularity assumptions, the number of elements in a vertex patch is limited, hence the family of master vertex patches is finite.

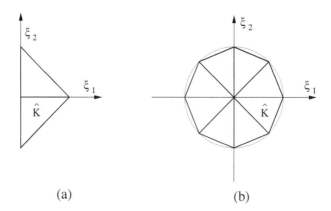

(a)	(b)

Figure 3.7: Examples of a master patch for (a) a boundary vertex node (case of a two elements patch) or an edge node, (b) an interior vertex node.

Let $T : \hat{S} \to S$ be a continuous union of affine transformations mapping master patch elements $\hat{K}$ onto patch S physical elements K,

$$\hat{K} \ni \xi \to A_K \xi + b_K \in K.$$

The shape regularity assumption implies that

$$\|A_K\|, \|A_K^{-1}\| \le C, \quad K \subset S,$$

for some $C > 0$.

Step 1: We first show that there exists a constant C such that, for any function $u \in H^1(S)$ with zero average, $(u, 1)_S = 0$, we have

$$\|u\|_{L^2(S)} \le C|u|_{H^1(S)}.$$

Let $\hat{u} = u \circ T$ be the pullback of u onto the master patch, and let c_0 denote the average value of $\hat{u}$ on $\hat{S}$. By Lemma 3.8, there exists then a constant $\hat{C}$, depending upon the master patch, such that

$$\|\hat{u} - c_0\|_{L^2(\hat{S})} \leq \hat{C}|\hat{u}|_{H^1(\hat{S})}.$$

As the number of master patches is limited, we can assume that constants $\hat{C}$ admit a common bound C. We now have

$$\|u\|_{L^2(S)}^2 \leq \|u\|_{L^2(S)}^2 + \|c_0\|_{L^2(S)}^2 = \|u - c_0\|_{L^2(S)}^2 \qquad ((u,1)_S = 0)$$

$$= \sum_{K \subset S} \|u - c_0\|_{L^2(K)}^2 \leq C \sum_{\hat{K} \subset \hat{S}} \|\hat{u} - c_0\|_{L^2(\hat{K})}^2$$

$$= C\|\hat{u} - c_0\|_{L^2(\hat{S})}^2 \leq C|\hat{u}|_{H^1(\hat{S})}^2$$

$$\leq C|u|_{H^1(S)}^2.$$

Step 2: We use the result now to prove the Lemma for the case $k = 0$. Let $\phi \in \mathcal{P}^{r-1}(S)$ be the unique polynomial[14] such that

$$(D^\alpha(u - \phi), 1)_S = 0, \qquad |\alpha| \leq r - 1.$$

Applying the preceding result to $D^\alpha(u - \phi)$, we get

$$\|u - \phi\|_{L^2(S)} \leq C \sum_{|\alpha|=1} \|D^\alpha(u - \phi)\|_{L^2(S)} \leq \cdots \leq C \sum_{|\alpha|=r} \|D^\alpha(u - \phi)\|_{L^2(S)} = C|u|_{H^r(S)}.$$

This gives

$$\|u - u_p\|_{L^2(S)} = \min_{\psi \in \mathcal{P}^p(S)} \|u - \psi\|_{L^2(S)} \leq \|u - \phi\|_{L^2(S)} \leq C|u|_{H^r(S)}.$$

Step 3: To prove the result for an arbitrary k we will need the *interpolation formula:*

$$|u|_{H^k(S)}^2 \leq C(\|u\|_{L^2(S)}^2 + |u|_{H^r(S)}^2).$$

The result formally follows immediately from Lemma 3.24; see Exercise 3.6.1. The delicate point is to demonstrate that one can find a constant C that would work for all patches. Let K be a triangle from a patch S, and let $\hat{K}$ be the corresponding master triangle in the master patch $\hat{S}$. We have

$$|u|_{H^k(K)}^2 \leq C|\hat{u}|_{H^k(\hat{K})}^2$$

$$\leq C(\|\hat{u}\|_{L^2(\hat{K})}^2 + |\hat{u}|_{H^r(\hat{K})}^2) \qquad \text{(Lemma 3.24)}$$

$$\leq C(\|u\|_{L^2(K)}^2 + |u|_{H^r(K)}^2).$$

The constant from Lemma 3.24 depends now upon the master triangle and, since the number of master triangles is limited, it allows for a uniform bound. Summing up over triangles in the patch, we get the desired result.

The second auxiliary result we need is the inverse estimate,

$$|\phi|_{H^k(S)} \leq C\|\phi\|_{L^2(S)}, \qquad \phi \in \mathcal{P}^p(S).$$

---

[14]Compare the proof of the Bramble–Hilbert Lemma.

This is proved using again the pullback maps and the finite-dimensionality argument,

$$|\phi|_{H^k(K)} \leq C|\hat{\phi}|_{H^k(\hat{K})} \leq C\|\hat{\phi}\|_{H^k(\hat{K})}$$

$$\leq C\|\hat{\phi}\|_{L^2(\hat{K})} \qquad \text{(norm equivalence in a finite-dimensional space)}$$

$$\leq C\|\phi\|_{L^2(K)} \,.$$

Again, it is critical that the norm equivalence argument is applied to a limited number of master triangles.

We proceed now with the proof.

Case: $p = r - 1, 0 \leq k \leq r$. We have

$$|u - u_p|^2_{H^k(S)} \leq C(\|u - u_p\|^2_{L^2(S)} + \underbrace{|u - u_p|^2_{H^r(\Omega)}}_{=|u|^2_{H^r(\Omega)}}) \qquad \text{(interpolation formula)}$$

$$\leq C|u|^2_{H^r(\Omega)} \qquad \qquad \text{(Step 2 result)} \,.$$

Case: $r < p + 1$. Let $\phi \in \mathcal{P}^{r-1}(S)$ be the L^2-projection of u onto $\mathcal{P}^{r-1}(S)$. We have

$$|u - u_p|_{H^k(S)} \leq |u - \phi|_{H^k(S)} + |\phi - u_p|_{H^k(S)} \,.$$

By the first case result, the first term is bounded by $|u|_{H^r(S)}$, and for the second term we have

$$|\phi - u_p|_{H^k(S)} \leq C\|\phi - u_p\|_{L^2(S)}$$

$$\leq C(\|u - \phi\|_{L^2(S)} + \|u - u_p\|_{L^2(S)}) \leq C|u|_{H^r(S)} \,. \qquad \square$$

Contrary to the previous lemma, the next one is simple.

Lemma 3.21. *Let K be an arbitrary simplicial element of order p. Then*

$$\|\phi\|_{L^\infty(K)} \leq C|K|^{-1/2}\|\phi\|_{L^2(K)}, \quad \phi \in \mathcal{P}^p(K) \,,$$

where C depends upon the master triangle and p, and $|K|$ is the measure (length, area, volume) of element K.

Proof.

$$\|\phi\|_{L^\infty(K)} = \|\hat{\phi}\|_{L^\infty(\hat{K})}$$

$$\leq C\|\hat{\phi}\|_{L^2(\hat{K})} \qquad \text{(equivalence of norms on a finite-dimensional space)}$$

$$\leq C|K|^{-1/2}\|\phi\|_{L^2(K)} \text{ (scaling)} \,. \qquad \square$$

Before we return to the proof of Clément's interpolation estimate, we note that mesh regularity conditions imply that the number of elements in a patch is bounded (we have already used this fact) and that

$$h_K = d(K) \leq d(S) \quad \text{and, conversely,} \quad d(S) \leq Ch_K \quad \text{for every element } K \text{ in patch } S \,.$$

Proof of Theorem 3.19. Let K be a triangle of order p with nodes $a_i, i = 1, \ldots, n := \dim(\mathcal{P}^p(K))$. Let $u \in H^r(\Omega)$, $r \leq p + 1$. Recalling the formula for the Clément interpolant,

$$\Pi_h u - u = \sum_{i=1}^n u_{i,p}(a_i)\phi_i - u = \underbrace{\sum_{i=1}^n u_{1,p}(a_i)\phi_i}_{=:u_{1,i}} - u + \sum_{i=1}^n (u_{i,p}(a_i) - u_{1,p}(a_i))\phi_i \,,$$

where $u_{i,p}$ is the L^2-projection of u onto $\mathcal{P}^p(S_i)$, with S_i denoting patch of elements corresponding to node a_i. The first term is bounded by the Lemma 3.20 result,

$$|u_{1,i} - u|_{H^k(K)} \leq |u_{1,i} - u|_{H^k(S)} \leq C\, d(S)^{r-k}\, |u|_{H^r(S)} \leq C\, h_K^{r-k}\, |u|_{H^r(S)}\,.$$

Next,

$$\begin{aligned}
\|u_{i,p} - u_{1,p}\|_{L^2(K)} &\leq \|u - u_{i,p}\|_{L^2(K)} + \|u - u_{1,p}\|_{L^2(K)} \\
&\leq C(d(S_i)^r |u|_{H^r(S_i)} + d(S_1)^r |u|_{H^r(S_1)}) \\
&\leq C h_K^r (|u|_{H^r(S_i)} + |u|_{H^r(S_1)})\,,
\end{aligned}$$

which, along with Lemma 3.21, implies

$$\|u_{i,p} - u_{1,p}\|_{L^\infty} |\phi_i|_{H^k(K)} \leq C\, h_K^{r-k}(|u|_{H^r(S_i)} + |u|_{H^r(S_1)})\,,$$

where we have used the bound

$$|\phi_i|_{H^k(K)} \leq C\, h_K^{-k}\, |K|^{1/2}\,.$$

Summing up the estimates, we get

$$|\Pi_h u - u|_{H^k(K)} \leq C\, h_K^{r-k} \sum_{i=1}^n |u|_{H^r(S_i)}\,.$$

Summing up over all elements in the mesh finishes the proof. □

Accounting for boundary conditions. If function u vanishes on part Γ_u of the domain boundary, we need the interpolant $\Pi_h u$ to vanish there as well. This is so far not the case, and we still need to modify the definition of interpolant to account for this behavior. We define the ultimate Clément interpolant by simply zeroing out the contributions from nodes on Γ_u. In other words, we perform the L^2-projections only for patches corresponding to nodes that are *not* on Γ_u,

$$\text{(modified)} \quad \tilde{\Pi}_h u = \sum_{a_i \notin \Gamma_u} u_{i,p}(a_i)\phi_i\,.$$

Theorem 3.22 (Clément, 1975). *Let $u \in H^r(\Omega)$, $r \leq p+1$, $u = 0$ on $\Gamma_u \subset \Gamma$. The following interpolation error estimates hold for the modified Clément interpolant:*

$$|u - \tilde{\Pi}_h u|_{H^k(\Omega)} \leq Ch^{r-k}|u|_{H^r(\Omega)}, \quad k = 0, 1, \ldots, r\,, \tag{3.6.82}$$

where constant C is independent of u.

The following is a result of the application of the Trace Theorem for a master element and scaling properties.

Lemma 3.23. *Let e be an edge of a triangular element K. There exists $C > 0$ such that*

$$h_K \|u\|_{L^2(e)}^2 \leq C\left[\|u\|_{L^2(K)}^2 + h_K^2 |u|_{H^1(K)}^2\right]$$

for every $u \in H^1(K)$.

Proof. Shape regularity assumptions imply that the length h_e of edge e and the element size h_K are of the same order. We then have

$$
\begin{aligned}
\|u\|_{L^2(e)}^2 = h_e\|\hat{u}\|_{L^2(\hat{e})}^2 &\leq h_K\|\hat{u}\|_{L^2(\hat{e})}^2 \\
&\leq Ch_K\left[\|\hat{u}\|_{L^2(\hat{K})}^2 + |\hat{u}|_{H^1(\hat{K})}^2\right] \qquad \text{(Trace Theorem)} \\
&\leq Ch_K\left[h_K^{-2}\|u\|_{L^2(K)}^2 + \|u\|_{H^1(K)}^2\right].
\end{aligned}
$$

Multiply both sides by h_K to finish the proof. $\square$

Proof of Theorem 3.22. It is sufficient to show that

$$
|\Pi_h u - \tilde{\Pi}_h u|_{H^k(K)}^2 \leq Ch_K^{2(r-k)}\sum_{i=1}^n |u|_{H^r(S_i)}^2, \qquad 0 \leq k \leq r.
$$

We have

$$
|\Pi_h u - \tilde{\Pi}_h u|_{H^k(K)} = \left|\sum_{a_i \in \Gamma_u} u_{i,p}(a_i)\phi_i\right|_{H^k(K)} \leq Ch_K^{-k}|K|^{1/2}\sum_{a_i \in \Gamma_u}|u_{i,p}(a_i)|. \qquad (3.6.83)
$$

Let $e \subset \partial K \cap \Gamma_u$, i.e., $u = 0$ on e, and let $a_i \in e$. Lemmas 3.23 and 3.20 imply that

$$
\begin{aligned}
h_K\|u_{i,p}\|_{L^2(e)}^2 = h_K\|u - u_{i,p}\|_{L^2(e)}^2 &\leq C\left[\|u - u_{i,p}\|_{L^2(S_i)}^2 + h_K^2|u - u_{i,p}|_{H^1(S_i)}^2\right] \qquad (3.6.84) \\
&\leq Ch_K^{2r}|u|_{H^r(S_i)}^2.
\end{aligned}
$$

Putting things together,

$$
\begin{aligned}
|\Pi_h u - \tilde{\Pi}_h u|_{H^k(K)}^2 &\leq Ch_K^{-2k}\underbrace{|K|}_{\approx h_K^2}\sum_{a_i \in \Gamma_u}|u_{i,p}|_{L^\infty(e)}^2 \qquad (3.6.83) \\
&\leq Ch_K^{-2k}\underbrace{|K|}_{\approx h_K^2}\sum_{a_i \in \Gamma_u}\underbrace{|e|^{-1}}_{h_K^{-1}}|u_{i,p}|_{L^2(e)}^2 \qquad \text{(Lemma 3.21)} \\
&\leq Ch_K^{2(r-k)}\sum_{a_i \in \Gamma_u}|u|_{H^r(S_i)}^2 \qquad (3.6.84)
\end{aligned}
$$

finishes the proof. $\square$

Exercises

3.6.1. Prove the following lemma. *Hint:* Look up the proof of Lemma 2.3.

Lemma 3.24. *Let $\Omega \subset \mathbb{R}^N$ be a bounded domain, and $k \geq 2$. There exists a constant C (depending upon the domain) such that*

$$
\|u\|_{H^k(\Omega)}^2 \leq C(\|u\|_{L^2(\Omega)}^2 + |u|_{H^k(\Omega)}^2)
$$

for every $u \in H^k(\Omega)$.

(5 points)

Chapter 4

Beyond Coercivity

In this chapter, we venture into general variational problems that may not be covered by the theory for coercive problems developed so far. We begin with the fundamental result of Ivo Babuška from 1971 that establishes a sufficient condition for the convergence of Petrov–Galerkin discretization for any well-posed variational problem, the famous *discrete inf-sup condition*. The next section covers the classical Mikhlin theory dealing with compact perturbations of Hermitian and coercive problems. This section is critical for those that work on vibrations and wave propagation problems. We present then the next fundamental topic—Franco Brezzi's theory for mixed problems (1973)—and discuss its equivalence with Babuška's Theorem. Finally we conclude with the fundamental result of Babuška, Kellogg and Pitkaranta showing that h-adaptivity may restore the optimal rate convergence dictated by the polynomial order of approximation alone, even for problems with singular solutions.

4.1 ▪ Babuška's Theorem

We begin with the fundamental theorem establishing the well-posedness theory for a general variational problem drawing from Banach's Closed Range Theorem.

Theorem 4.1 (Babuška–Nečas Theorem). *Consider the standard abstract variational problem,*

$$\begin{cases} u \in U, \\ b(u,v) = l(v) \quad \forall v \in V, \end{cases} \tag{4.1.1}$$

where U, V are Hilbert (trial and test) spaces, $b(u,v)$ is a continuous bilinear (sesquilinear) form, and $l \in V'$ is a continuous linear (antilinear) form on test space V. Additionally, assume that b satisfies the inf-sup condition,

$$\inf_{u \in U, u \neq 0} \sup_{v \in V, v \neq 0} \frac{|b(u,v)|}{\|u\|_U \|v\|_V} \geq \gamma > 0,$$

and l satisfies the compatibility condition,

$$l(v) = 0 \quad \forall v \in V_0 := \{v \in V : b(u,v) = 0 \quad \forall u \in U\}.$$

There exists then a unique solution u to the variational problem and it satisfies the stability estimate:

$$\|u\|_U \leq \frac{1}{\gamma} \|l\|_{V'}.$$

Proof. The result is a reinterpretation of the Banach Closed Range Theorem; see [66, p. 518] for details. $\square$

If $P : U \to U$ is an orthogonal projection in a Hilbert space U, then so is $I - P$, and both have a unit norm. Consequently, $\|I - P\| = \|P\|$ (=1). It turns out that the result holds for *any* linear (oblique) projection P *defined on a Hilbert space* as well.

Lemma 4.2 (Del Pasqua, Ljance, and Kato [69]).   *Let $U, (\cdot, \cdot)$ be a Hilbert space, and let $P : U \to U$ be a linear projection, i.e., $P^2 = P$. Then*

$$\|I - P\| = \|P\| \,.$$

Proof. Let $X = \mathcal{R}(P)$ and $Y = \mathcal{N}(P)$. It is well known that $U = X \oplus Y$. Pick an arbitrary unit vector $u \in U$. Let $u = x + y$, $x \in X$, $y \in Y$ be the unique decomposition of u. By the properties of a scalar product,

$$1 = \|u\|^2 = (x + y, x + y) = \|x\|^2 + \|y\|^2 + 2\mathrm{Re}(x, y) \,.$$

Consider now a "symmetric image" w of u (see Fig. 4.1),

$$w = \bar{x} + \bar{y}, \quad \bar{x} = \|y\| \frac{x}{\|x\|}, \quad \bar{y} = \|x\| \frac{y}{\|y\|} \,.$$

Vector w has a unit length as well. Indeed,

$$\|w\|^2 = (\bar{x}+\bar{y}, \bar{x}+\bar{y}) = \|\bar{x}\|^2 + \|\bar{y}\|^2 + 2\mathrm{Re}(\bar{x}, \bar{y}) = \|y\|^2 + \|x\|^2 + 2\mathrm{Re}\left(\frac{\|y\|}{\|x\|}\frac{\|x\|}{\|y\|}(x, y)\right) = 1 \,.$$

We now have

$$\|Pu\| = \|x\| = \|\bar{y}\| = \|(I - P)w\| \leq \|I - P\| \, \|w\| = \|I - P\| \,.$$

Taking the supremum over $\|u\| = 1$ finishes the proof. $\square$

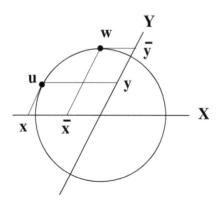

Figure 4.1: Illustration of the proof of the Del Pasqua–Ljance–Kato Lemma.

Remark 4.3. The geometrical structure of Hilbert space in Lemma 4.2 is critical. The result is no longer true for projections in a Banach space U, even when U is reflexive; compare Exercise 4.1.1.

Theorem 4.4 (Babuška Theorem [4, 71]). *Consider the Petrov–Galerkin discretization of variational problem (4.1.1),*

$$\begin{cases} u_h \in U_h \,, \\ b(u_h, v_h) = l(v_h) \quad \forall v_h \in V_h \,, \end{cases} \tag{4.1.2}$$

where $U_h \subset U$, $V_h \subset V$ are discrete trial and test spaces, and $\dim U_h = \dim V_h < \infty$. Assume that form b and the discrete spaces satisfy the discrete version of the inf-sup condition:

$$\inf_{u_h \in U_h, u_h \neq 0} \sup_{v_h \in V, v_h \neq 0} \frac{|b(u_h, v_h)|}{\|u_h\|_U \, \|v_h\|_V} =: \gamma_h > 0 \,.$$

There exists then a unique (discrete) solution u_h to variational problem (4.1.2) which satisfies the stability estimate:

$$\|u_h\|_U \leq \frac{1}{\gamma_h} \|l\|_{V_h'} \,.$$

Additionally, we have the estimate

$$\|u - u_h\| \leq \frac{M}{\gamma_h} \inf_{w_h \in U_h} \|u - w_h\|_U \,,$$

where M is the continuity constant for form b.

Proof. The stability result is a direct consequence of the Babuška–Nečas Theorem as the discrete variational problem is simply a particular case of the general case. Notice that no compatibility condition is needed on the discrete level; the Galerkin stiffness matrix and its transpose have the same rank. The proof of the error estimate (4.1.3) begins with an observation that the Petrov–Galerkin discretization executes a linear projection $P_h : U \to U_h$, $P_h u = u_h$,

$$b(P_h u - u, v_h) = 0 \quad \forall v_h \in V_h \,.$$

The stability estimate implies an estimate on the norm of the projection,

$$\|P_h u\|_U = \|u_h\| \leq \frac{1}{\gamma_h} \|l\|_{V_h'} = \frac{1}{\gamma_h} \sup_{\|v_h\|=1} |b(u, v_h)| \leq \frac{M}{\gamma_h} \|u\|_U \,.$$

We then have

$$\begin{aligned} \|u - u_h\|_U &= \|(I - P_h)u\|_U & \text{(definition of } P_h) \\ &= \|(I - P_h)(u - w_h)\|_U & (P_h w_h = w_h \quad \forall w_h \in U_h) \\ &\leq \|I - P_h\| \, \|u - w_h\| \\ &= \|P_h\| \, \|u - w_h\| & \text{(Lemma 4.2)} \\ &\leq \frac{M}{\gamma_h} \|u - w_h\| & (\|P_h\| \leq \frac{M}{\gamma_h}) \,, \end{aligned}$$

and we conclude the proof by taking the infimum over $w_h \in U_h$. $\quad\square$

If γ_h admit a positive lower bound, i.e., a uniform discrete inf-sup condition holds,

$$\inf_h \gamma_h =: \gamma_0 > 0 \,,$$

then

$$\underbrace{\|u - u_h\|}_{\text{approximation error}} \leq \underbrace{\frac{M}{\gamma_0}}_{\text{stability constant}} \underbrace{\inf_{w_h \in U_h} \|u - w_h\|_U}_{\text{best approximation error}}, \tag{4.1.3}$$

i.e., the actual and the best approximation errors must converge at the same rate. The result has coined the famous phrase:

(Uniform) discrete stability and approximability imply convergence.

It is not an exaggeration to say that the entire numerical analysis for linear problems hinges on the Babuška Theorem. The result, unfortunately, is not constructive. It tells us what we should have to ensure the stability and convergence for the Galerkin method, but it gives no hint how to select the spaces to guarantee the discrete inf-sup condition. However, the result underlines the different criteria for selection of spaces: the trial space choice controls the approximability error, whereas the test space controls the stability. In the case of $U = V$, we may choose $V_h = U_h$ (Bubnov–Galerkin method), but the control of stability becomes then incidental, and we may not have it.

Remark 4.5. The presented proof is due to Xu and Zikatanov [71]. The original proof of Ivo Babuška did not use Lemma 4.2. It is much simpler, and it holds for a class of more general variational formulations set up in (reflexive) Banach spaces, but it provides a suboptimal stability constant in the Hilbert space setting. We record it for completeness. Let $w_h \in U_h$ be arbitrary. We have

$$\|u_h - w_h\|_U \leq \gamma_h^{-1} \sup_{v_h \in V_h} \frac{|b(u_h - w_h, v_h)|}{\|v_h\|_V} \qquad \text{(discrete inf-sup condition)}$$

$$\leq \gamma_h^{-1} \sup_{v_h \in V_h} \frac{|b(u - w_h, v_h)|}{\|v_h\|_V} \qquad \text{(Galerkin orthogonality)}$$

$$\leq \frac{M}{\gamma_h} \|u - w_h\|_U \qquad \text{(continuity of form b)}.$$

By the triangle inequality,

$$\|u - u_h\|_U \leq \|u - w_h\|_U + \|w_h - u_h\|_U \leq \underbrace{\left(1 + \frac{M}{\gamma_h}\right)}_{\text{stability constant}} \|u - w_h\|_U.$$

Taking infimum with respect to w_h finishes the proof. For sharper estimates of the stability constant in the context of Banach spaces, see [60].

Exercises

4.1.1. Construct a counterexample for Lemma 4.2 if U is only Banach. *Hint:* Consider $U = \mathbb{R}^2$ equipped with an L^1-norm. (5 points)

4.1.2. Discrete inf-sup constant. Let $e_i \in U_h$ and $g_j \in V_h$, $i, j = 1, \ldots, \dim U_h = \dim V_h$, be specific basis functions for the discrete trial and test space. Introduce the Galerkin stiffness matrix and the corresponding Gram matrices for the norms

$$B_{ji} := b(e_i, g_j), \quad (G_U)_{ji} := (e_i, e_j)_U, \quad (G_V)_{ji} := (g_i, g_j)_V.$$

Derive an explicit formula for the discrete inf-sup constant γ_h in terms of matrices $B, G_U,$ G_V. Can you compute it using standard (iterative) algorithms for computing eigenvalues? (5 points)

4.2 ▪ Asymptotic Stability

Solomon Mikhlin published his theory on asymptotic stability in 1959, five years before Cea's Lemma and, from the historical perspective, the results discussed in this section should follow the Ritz theory. It is easier, though, to discuss them being familiar first with Babuška's Theorem, hence the order of presentation.

Consider a class of variational problems of the form

$$\begin{cases} u \in V, \\ \underbrace{a(u,v) + c(u,v)}_{=:b(u,v)} = l(v), \quad v \in V, \end{cases} \tag{4.2.4}$$

where V is a Hilbert space, sesquilinear form $a(u,v)$ is Hermitian and coercive,

$$a(u,v) = \overline{a(v,u)}, \quad u, v \in V, \qquad a(u,u) \geq \alpha \|u\|_V^2, \quad u \in V, \alpha > 0,$$

$l \in V'$, and form $c(u,v)$ is *compact*. From many possible definitions of a compact form, we choose the one as follows. Form $c(u,v)$ is said to be *compact* iff

$$u_n \rightharpoonup u \quad \Rightarrow \quad \sup_{\|v\|_V \leq 1} |c(u_n - u, v)| \to 0. \tag{4.2.5}$$

Remark 4.6. Let $c(u,v)$ be compact. Then

$$c(u_n - u, u_n - u) \to 0 \quad \text{if} \quad u_n \rightharpoonup u.$$

Indeed, you need to recall only that weak convergence implies boundedness.

Example 4.7. Assume that the energy space V is compactly embedded in another Hilbert space H,

$$V \overset{c}{\hookrightarrow} H,$$

i.e.,

$$u_n \rightharpoonup u \text{ in } V \quad \Rightarrow \quad u_n \to u \text{ in } H,$$

and

$$|c(u,v)| \leq C \|u\|_H \|v\|_V,$$

i.e., $c(u,v)$ is continuous on weaker space $H \times V$. It follows then immediately from the definition that $c(u,v)$ is compact. A specific example of such a scenario will be the Helmholtz problem:

$$\begin{cases} u \in H_0^1(\Omega), \\ (\nabla u, \nabla v) - \omega^2(u,v) = (f,v), \quad v \in H_0^1(\Omega). \end{cases}$$

Above, as usual, $(\cdot, \cdot)$ denotes the L^2-inner product.

We shall also make the following assumption.

Density assumption.

$$\forall\, v \in V \quad \exists\, v_h \in V_h \; : \; \|v_h - v\|_V \to 0 \quad \text{as } h \to 0. \tag{4.2.6}$$

In other words, $\bigcup_h V_h$ is dense in V.

Theorem 4.8 (Mikhlin [58]). *Consider problem* (4.2.4) *and assume that density assumption* (4.2.6) *holds. Assume additionally that operator B corresponding to sesquilinear form $b(u,v)$ is injective. Then*

$$\exists\, h_0 \quad \exists\, \gamma_0 \quad \forall\, h < h_0 \quad (\forall\, u_h \in V_h) \quad \sup_{v_h \in V_h} \frac{|b(u_h, v_h)|}{\|v_h\|} \geq \gamma_0 \|u_h\|. \tag{4.2.7}$$

In other words, the problem is asymptotically stable.

Proof. Assume, to the contrary, that

$$\forall\, h_0 \quad \forall\, \gamma_0 \quad \exists\, h < h_0 \quad \exists\, u_h \in V_h \quad \sup_{v_h \in V_h} \frac{|b(u_h, v_h)|}{\|v_h\|} < \gamma_0 \|u_h\|.$$

Set $\gamma_n = 1/n$, $h_0 = 1/n$ to conclude existence of a sequence $h_n < 1/n$, and the corresponding sequence of unit vectors $\|u_{h_n}\| = 1$ such that

$$\sup_{v_h \in V_h} \frac{|b(u_h, v_h)|}{\|v_h\|} \leq \frac{1}{n}.$$

Recall then that, in a Hilbert space, every bounded sequence has a weakly convergent subsequence. Replace the original sequence with the subsequence. We have thus $u_{h_n} \rightharpoonup u_0$. We claim that the sequence u_{h_n} converges to u_0 actually *strongly*. Indeed, coercivity of form $a(u,v)$ implies

$$
\begin{aligned}
\alpha \|u_0 - u_{h_n}\|^2 &\leq a(u_0 - u_{h_n}, u_0 - u_{h_n}) \\
&= b(u_0 - u_{h_n}, u_0 - u_{h_n}) - c(u_0 - u_{h_n}, u_0 - u_{h_n}) \\
&= b(u_0, u_0 - u_{h_n}) - b(u_{h_n}, u_0 - u_{h_n}) - c(u_0 - u_{h_n}, u_0 - u_{h_n}).
\end{aligned}
$$

The first term converges to zero by the definition of weak convergence ($b(u_0, \cdot) \in V'$), and the third one converges to zero by Remark 4.6. It remains to show that the second term converges to zero as well.

By the density assumption, we can select a sequence $w_{h_n} \to u_0$. We have then

$$|b(u_{h_n}, u_0 - u_{h_n})| \leq |b(u_{h_n}, u_0 - w_{h_n})| + |b(u_{h_n}, w_{h_n} - u_{h_n})|$$

$$\leq M \underbrace{\|u_{h_n}\|}_{\text{bounded}} \underbrace{\|u_0 - w_{h_n}\|}_{\to 0} + \underbrace{\sup_{v_{h_n}} \frac{|b(u_{h_n}, v_{h_n})|}{\|v_{h_n}\|}}_{\leq \frac{1}{n} \to 0} \underbrace{\|w_{h_n} - u_{h_n}\|}_{\text{bounded}}.$$

Strong convergence of u_{h_n} to u_0 implies that $1 = \|u_{h_n}\| \to \|u_0\|$, so $\|u_0\| = 1$ and, therefore, $u_0 \neq 0$. Consider then arbitrary v and a sequence v_{h_n} converging strongly to v. By continuity of form $b(u,v)$,

$$b(u,v) = \lim_{n \to \infty} b(u_{h_n}, v_{h_n}).$$

However,

$$|b(u_{h_n}, v_{h_n})| \leq \sup_{v_h \in V_h} \frac{|b(u_{h_n}, v_h)|}{\|v_h\|} \|v_{h_n}\| \leq \frac{1}{n} \underbrace{\|v_{h_n}\|}_{\text{bounded}} \to 0.$$

Consequently,

$$b(u_0, v) = 0 \quad \forall v \in V,$$

which contradicts the uniqueness of the solution (injectivity of B).   □

The Mikhlin result tells us that, asymptotically, the Bubnov–Galerkin method is stable. It does not shed light on how small (or large) the parameter h_0 should be. The following simple but representative example provides some intuition.

Vibrations. A model problem.   Consider an abstract variational problem,

$$\begin{cases} u \in V, \\ a(u, v) - \omega^2 m(u, v) = m(f, v), \quad v \in V, \end{cases} \tag{4.2.8}$$

where V is an energy space, Hermitian, coercive form $a(u, v)$ represents the elastic energy, Hermitian and positive definite form $m(u, v)$ represents the mass, ω is the forcing frequency, and f is a force per unit mass. Consider the variational eigenproblem,

$$\begin{cases} e_i \in V, \\ a(e_i, v) = \omega_i^2 m(e_i, v), \quad v \in V. \end{cases} \tag{4.2.9}$$

If mass represents a compact perturbation of energy (case of a bounded domain), there exists an infinite number of eigenpairs (ω_i^2, e_i) with real and positive eigenvalues $\omega_i^2 \to \infty$, and eigenvectors e_i providing an orthogonal (in terms of both mass and energy) basis for V. We equip the energy space with the energy norm

$$\|u\|^2 := a(u, u)$$

and assume that the eigenvectors have been normalized with mass, i.e.,

$$m(e_i, e_i) = 1, \quad i = 1, 2, \ldots.$$

We will now compute explicitly the corresponding inf-sup constant γ and continuity constant M in terms of the eigenvalues ω_i^2. Let $u, v \in V$ and

$$u = \sum_{i=1}^{\infty} u_i e_i, \quad v = \sum_{j=1}^{\infty} v_j e_j$$

be the corresponding spectral representations. We can represent the energy norm in terms of spectral components,

$$\|u\|^2 = \sum_{i=1}^{\infty} \omega_i^2 u_i^2, \quad \|v\|^2 = \sum_{j=1}^{\infty} \omega_j^2 v_j^2.$$

We have

$$\sup_v \frac{|a(u, v) - \omega^2 m(u, v)|}{\|v\|} = \|a(u, \cdot) - \omega^2 m(u, \cdot)\|_{V'} = \|v\|,$$

where $v = R_V^{-1}(a(u, \cdot) - \omega^2 m(u, \cdot))$, with R_V denoting the Riesz operator. We get

$$\omega_i^2 v_i = (\omega_i^2 - \omega^2) u_i$$

and

$$\|v\|^2 = \sum_{i=1}^{\infty} \omega_i^2 |v_i|^2 = \sum_{i=1}^{\infty} \omega_i^2 \left(\frac{\omega_i^2 - \omega^2}{\omega_i^2} \right)^2 |u_i|^2 .$$

The inf-sup constant γ satisfies

$$\sum_{i=1}^{\infty} \omega_i^2 \left(\frac{\omega_i^2 - \omega^2}{\omega_i^2} \right)^2 |u_i|^2 \geq \gamma^2 \sum_{i=1}^{\infty} \omega_i^2 |u_i|^2 .$$

Comparing coefficients, we get

$$\gamma = \min_i \frac{\omega_i^2 - \omega^2}{\omega_i^2} .$$

Notice that, despite the infinite number of spectral components, the minimum is actually attained (explain why). Concerning the continuity constant, we have

$$|b(u,v)| = \left| \sum_i (\omega_i^2 - \omega^2) u_i \overline{v_i} \right| = \left| \sum_i \frac{\omega_i^2 - \omega^2}{\omega_i^2} \omega_i u_i \, \omega_i \overline{v_i} \right| \leq \max_i \left| 1 - \left(\frac{\omega}{\omega_i} \right)^2 \right| \|u\| \, \|v\| .$$

We see that the continuity constant M is of order ω^2, whereas the inf-sup constant $\gamma = 0$ if the forcing frequency ω matches one of the eigenfrequencies ω_i (resonance). The same reasoning can be repeated for the discrete problem using discrete eigenpairs $(\omega_{h,i}^2, e_{h,i})$. This leads to the analogous formula for the discrete inf-sup constant γ_h,

$$\gamma_h^{-1} = \frac{1}{\min_i |1 - (\frac{\omega}{\omega_{h,i}})^2|} .$$

It is well known that the discrete eigenvalues $\omega_{h,i}^2$ converge monotonically from above to the corresponding exact eigenvalues ω_i^2; compare Exercise 4.2.3. Imagine that the forcing frequency ω happens to be in between the exact eigenfrequency ω_i and, for some mesh, the corresponding discrete eigenfrequency $\omega_{h,i}$; compare Fig. 4.2. As you keep refining (uniformly) the mesh, $\omega_{h,i}$, marching toward ω_i, has to migrate over the forcing frequency ω. It may even hit ω and the discrete problem will then become unstable (ill-posed). Or it can get so close to ω that the round-off error will make the discrete problem effectively singular. The moral of the story is that only once $\omega_{h,i}$ migrates to the left side of ω is the danger of resonance (or quasi-resonance) gone. From now on, the global refinements will lead to stable discrete problems. Of course, the criterion of being on the correct side of the forcing frequency must be satisfied for all eigenfrequencies. In short, the stability is related to the convergence of eigenfrequencies that are close to the forcing frequency.

You can also see that, eventually, the discrete inf-sup constant converges to the exact one. Asymptotically, stability of the discrete problem reflects stability of its continuous counterpart.

For problems with damping, we lose the orthogonality structure and such a characterization becomes impossible although the inf-sup constant can still be represented in terms of the singular values of the operator, and the discrete inf-sup constant still converges to the exact one; see [23].

Remark 4.9. Being in the preasymptotic range is accompanied by a flip in a spectral component of the solution. Spectral components u_i of the solution to (4.2.8) are given by the formula

$$u_i = \frac{f_i}{\omega_i^2 - \omega^2} .$$

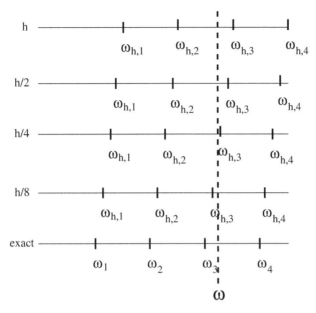

Figure 4.2: Migration of discrete eigenvalues.

The same formula holds on the discrete level,

$$u_{h,i} = \frac{f_{h,i}}{\omega_{h,i}^2 - \omega^2} .$$

If the sign of factor $\omega_{h,i}^2 - \omega^2$ is different from that of $\omega_i^2 - \omega^2$, the component will be "flipped." If you are in a quasi-resonant mode, this may be the largest component of the solution. Do not look then for a bug in your code as I did many years ago, losing several days before I understood the phenomenon.

Asymptotic optimality of the Galerkin method [37]. We will make just two assumptions: (a) the method converges in the energy norm implied by the leading Hermitian term,

$$\|u\|_E^2 := a(u, u) ,$$

and (b) the method converges with a faster rate in the weaker norm $\| \cdot \|_H$ controlling the continuity of compact perturbation term $c(u, v)$,

$$|c(u, v)| \le C \|u\|_H \|v\|_E .$$

We then have

$$\|u - u_h\|_E^2 - C \|u - u_h\|_H \|u - u_h\|_E$$
$$\le |b(u - u_h, u - u_h)|$$
$$= |b(u - u_h, u - w_h)| \quad \text{(Galerkin orthogonality)}$$
$$= a(u - u_h, u - w_h) + |c(u - u_h, u - w_h)|$$
$$\le \|u - u_h\|_E \|u - w_h\|_E + C \|u - u_h\|_H \|u - w_h\|_E ,$$

or

$$\|u - u_h\|_E \leq \frac{\|u - u_h\|_E + C\|u - u_h\|_H}{\|u - u_h\|_E - C\|u - u_h\|_H} \inf_{w_h \in U_h} \|u - w_h\|_E\,.$$

Due to the faster convergence in the weaker norm, the fraction on the right-hand side converges asymptotically to unity. Hence, asymptotically, the Galerkin error converges to the best approximation error.

Relation with the compact perturbation of identity. In this section, we relate Mikhlin's problem with the classical theory of compact perturbations of identity in a Hilbert space [66, Section 5.20]. The variational problem can be rewritten in the operator form,

$$Bu = Au + Cu = l\,, \tag{4.2.10}$$

where operators $A, B, C : V \to V'$ correspond to sesquilinear forms a, b, c. Compactness of form $c(u, v)$ is equivalent to compactness of operator C. We now equip space V with the equivalent energy norm

$$\|v\|_E^2 = a(v, v)\,.$$

Form $a(u, v)$ then becomes the inner product on V, and operator A becomes the corresponding Riesz operator. Applying A^{-1} to both sides of operator equation (4.2.10), we can replace the problem with an equivalent problem of the form

$$(I + A^{-1}C)u = A^{-1}l\,.$$

Operator $K := A^{-1}C$, as a composition of compact operator C and continuous operator A^{-1}, is compact. Equivalently,

$$\begin{cases} Ku \in V\,, \\ a(Ku, v) = c(u, v), \quad v \in V\,. \end{cases}$$

The classical Fredholm alternative for operators of the second kind, $I + K$ where K is compact, implies that injectivity of $I + K$ implies its boundedness below and invertibility on the whole space. Since

$$\sup_{\|v\| \leq 1} \frac{|\langle (A + C)u, v \rangle|}{\|v\|_E} = \|A^{-1}\langle (A + C)u, \cdot \rangle\|_E = \|(I + K)u\|_E\,,$$

the inf-sup constant corresponding to the energy norm can be reinterpreted as

$$\gamma = \inf_{\|u\|_E = 1} \|(I + K)u\|_E\,.$$

Let $V_h \subset V$ now be a finite-dimensional subspace of V. The discrete inf-sup constant is characterized in the same way,

$$\gamma_h = \inf_{\|u_h\|_E = 1} \|(I + K_h)u_h\|_E\,,$$

where discrete operator $K_h : V_h \to V_h$ is defined by

$$\begin{cases} K_h u_h \in V\,, \\ a(K_h u_h, v_h) = c(u_h, v_h), \quad v_h \in V_h\,. \end{cases}$$

The following theorem summarizes fundamental relations between γ, γ_h and operators K, K_h.

Theorem 4.10 ([23]). *Let*

$$\|K - K_h\| := \sup_{\|v_h\|_E \leq 1, \, v_h \in V_h} \|(K - K_h)v_h\|_E \,.$$

The following properties hold:

(i)
$$\|K - K_h\| \to 0 \quad as \ h \to 0 \,, \tag{4.2.11}$$

(ii)
$$\gamma_h \geq \gamma - \|K - K_h\| \,, \tag{4.2.12}$$

(iii)
$$\gamma_h \to \gamma \quad as \ h \to 0 \,. \tag{4.2.13}$$

Proof.

(i) We have the orthogonality condition,

$$a(K u_h - K_h u_h, v_h) = 0 \,, \quad v_h \in V_h \,.$$

Assume, to the contrary, that (4.2.11) does not hold, i.e.,

$$\exists \epsilon \quad \forall h_0 \quad \exists h < h_0 \quad \exists \|u_h\|_E = 1, \quad \|(K - K_h)u_h\| \geq \epsilon \,.$$

Selecting $h_0 = 1/n$, we obtain a sequence u_{h_n} such that

$$\|u_{h_n}\|_E = 1, \quad h_n \to 0, \quad \|(K - K_{h_n})u_{h_n}\| \geq \epsilon \,.$$

By the weak compactness argument, replacing the sequence with some subsequence, we can additionally assume that u_{h_n} converges weakly to some u_0. By the orthogonality condition above, we have

$$
\begin{aligned}
\|(K - K_{h_n})u_{h_n}\|_E^2 &= a((K - K_{h_n})u_{h_n}, (K - K_{h_n})u_{h_n}) \\
&= a((K - K_{h_n})u_{h_n}, K u_{h_n} - v_{h_n}) \\
&\leq \|(K - K_{h_n})u_{h_n}\|_E \|K u_{h_n} - v_{h_n}\|_E
\end{aligned}
$$

for arbitrary $v_{h_n} \in V_{h_n}$. Hence,

$$\|(K - K_{h_n})u_{h_n}\|_E \leq \|K u_{h_n} - v_{h_n}\|_E \,.$$

But, by compactness of K, $K u_{h_n}$ converges strongly to $K u_0$ and, by density assumption (4.2.6), one can select a sequence v_{h_n} converging to $K u_0$. Consequently, the right-hand side converges to zero, and therefore the left-hand side does as well, a contradiction.

(ii) We have
$$\|(I + K)u_h\|_E \leq \|(I + K_h)u_h\|_E + \|K - K_h\| \|u_h\|_E \,.$$

Consequently,

$$
\begin{aligned}
\inf_{\|u\|_E = 1, u \in V} \|(I + K)u\|_E &\leq \inf_{\|u_h\|_E = 1, u_h \in V_h} \|(I + K)u_h\|_E \\
&\leq \inf_{\|u_h\|_E = 1, u_h \in V_h} \|(I + K_h)u_h\|_E + \|K - K_h\| \,.
\end{aligned}
$$

(iii)
$$\gamma_h = \inf_{\|u_h\|_E=1} \|(I + K_h)u_h\|_E \leq \inf_{\|u_h\|_E=1} \|(I + K)u_h\|_E + \|K - K_h\| .$$

But
$$\inf_{\|u_h\|_E=1} \|(I + K)u_h\|_E \to \inf_{\|u\|_E=1} \|(I + K)u\|_E = \gamma$$

and, therefore,
$$\limsup_{h\to 0} \gamma_h \leq \gamma .$$

which, along with (4.2.12), finishes the proof.   □

Remark 4.11. Condition (ii) implies that attaining the asymptotic stability region depends upon the value of inf-sup constant γ and the rate of convergence of $\|K - K_h\|$ to zero. Condition (iii) indicates that, asymptotically in h, the discrete inf-sup constant cannot be better than the continuous one. In other words, one cannot expect to obtain a well-conditioned approximate problem from an ill-conditioned continuous one.

Exercises

4.2.1. Spectral decomposition. Let $a(u, v)$ be a continuous and coercive Hermitian form on a Hilbert space V, and let $c(u, v)$ be a positive definite, Hermitian, and compact form on the same space V. Use the spectral theory for self-adjoint compact operators in a Hilbert space [66, Section 6.10] to conclude the existence of eigenpairs (λ_i, e_i),

$$\begin{cases} e_i \in V , \\ a(e_i, v) = \lambda_i\, c(e_i, v), \quad v \in V , \end{cases}$$

$i = 1, 2, \ldots$, such that

$$0 < \lambda_1 \leq \lambda_2 \leq \cdots \leq \lambda_n \to \infty \quad \text{as } n \to \infty$$

and

$$c(e_i, e_j) = \delta_{ij}, \quad a(e_i, e_j) = \lambda_i \delta_{ij} \text{ (no summation)} .$$

Moreover, the following representations hold:

$$a(u, u) = \sum_{i=1}^{\infty} \lambda_i |u_i|^2, \quad c(u, u) = \sum_{i=1}^{\infty} |u_i|^2 , \tag{4.2.14}$$

where

$$u = \sum_{i=1}^{\infty} u_i e_i := \lim_{n\to\infty} \sum_{i=1}^{n} u_i e_i .$$

(5 points)

4.2.2. Variational principles for generalized eigenvalues.   Consider the scenario from Exercise 4.2.1. Prove the following variational principles for generalized eigenpairs (λ_i, e_i).

(i) *Rayleigh quotient:*

$$\min_{u\in V} \frac{a(u, u)}{c(u, u)} = \frac{a(e_1, e_1)}{c(e_1, e_1)} = \lambda_1 .$$

(ii) *Generalized Rayleigh quotient:*

$$\min_{\substack{u \in V \\ c(u, e_i),\, i = 1, \ldots, n-1}} \frac{a(u, u)}{c(u, u)} = \frac{a(e_n, e_n)}{c(e_n, e_n)} = \lambda_n \,.$$

(iii) *Min-max principle:*

$$\min_{\substack{V_n \subset V \\ \dim V_n = n}} \max_{u \in V_n} \frac{a(u, u)}{c(u, u)} = \max_{u \in \mathrm{span}\{e_1, \ldots, e_n\}} \frac{a(u, u)}{c(u, u)} = \frac{a(e_n, e_n)}{c(e_n, e_n)} = \lambda_n \,.$$

(5 points)

4.2.3. **Convergence of eigenvalues.** Consider the scenario from Exercise 4.2.1. Let $V_h \subset V$ be a finite-dimensional subspace, $\dim V_h = N_h$. Consider the approximate eigenvalue problem,

$$\begin{cases} e_{h,i} \in V_h \,, \\ a(e_{h,i}, v_h) = \lambda_{h,i}\, c(e_{h,i}, v_h), \quad v_h \in V_h \,. \end{cases}$$

Use the min-max principle from Exercise 4.2.2 to prove that

$$\lambda_i \le \lambda_{h,i}, \quad i = 1, \ldots, N_h \,.$$

Argue why the approximability condition,

$$\forall\, v \in V \quad \exists v_h \in V_h, \quad \|v - v_h\|_V \to 0 \quad \text{as } h \to 0 \,,$$

implies $\lambda_{h,i} \to \lambda_i$, $i = 1, 2 \ldots$. (5 points)

4.3 ▪ Mixed Problems

In this section, we review the famous theory of Franco Brezzi for mixed problems and relate it to the Babuška–Nečas and Babuška Theorems.

Constrained minimization problems. Consider the standard (potential energy) functional defined on a Hilbert space V,

$$J(v) = \frac{1}{2} a(v, v) - \Re f(v), \quad v \in V \,,$$

where $f \in V'$, and $a(u, v)$ is a sesquilinear, Hermitian, coercive form on $V \times V$,

$$a(v, v) \ge \alpha \|v\|_V^2, \quad v \in V, \quad \alpha > 0 \,.$$

Let Q be another Hilbert space, and let $b(v, q)$, $v \in V, q \in Q$, denote another sesquilinear form. Consider a *constrained minimization problem*,

$$\inf_{v \in V_g} J(v) \,,$$

where

$$V_g := \{v \in V \,:\, b(v, q) = g(q) \quad \forall q \in Q\} \,,$$

with a given[15] $g \in Q'$. Recall that, for a complex setting,

$$b(v, q) - g(q) = 0, \quad q \in Q \qquad \Longleftrightarrow \qquad \Re(b(v, q) - g(q)) = 0, \quad q \in Q.$$

In order to derive necessary conditions for the minimizer, introduce the Lagrangian

$$L(v, q) := J(v) + \Re(b(v, q) - g(q))$$

and differentiate it with respect to v and q to obtain

$$\begin{cases} u \in V, \, p \in Q, \\ \Re\left(a(u, v) + b^*(p, v) - f(v)\right) = 0, \quad v \in V, \\ \Re\left(b(u, q) - g(q)\right) = 0, \quad q \in Q, \end{cases}$$

or, equivalently,

$$\begin{cases} u \in V, \, p \in Q, \\ a(u, v) + b^*(p, v) = f(v), \quad v \in V, \\ b(u, q) = g(q), \quad q \in Q, \end{cases} \qquad (4.3.15)$$

with $b^*(p, v) = \overline{b(v, p)}$. Problem (4.3.15) is identified as a *mixed problem* to be solved for the minimizer u and the Lagrange multiplier p.

Eventually, we extend our interest to a larger class of mixed problems where the form $a(u, v)$ may be neither Hermitian nor coercive.

The mixed problem can be cast into the standard variational setting by introducing the group variables,

$$\mathsf{u} := (u, p) \in V \times Q, \quad \mathsf{v} := (v, q) \in V \times Q,$$

and a "big" sesquilinear form,

$$\mathsf{b}(\mathsf{u}, \mathsf{v}) := a(u, v) + b^*(p, v) + b(u, q) = a(u, v) + \overline{b(v, p)} + b(u, q).$$

Mixed problem (4.3.15) is then equivalent to

$$\begin{cases} \mathsf{u} \in V \times Q, \\ \mathsf{b}(\mathsf{u}, \mathsf{v}) = \mathsf{l}(\mathsf{v}), \quad \mathsf{v} \in V \times Q, \end{cases} \qquad (4.3.16)$$

where $\mathsf{l}(\mathsf{v}) := f(v) + g(q)$.

Babuška $\Rightarrow$ Brezzi. Our main tool in deriving the famous Brezzi's conditions [11, 46, 7] will be the following fundamental property of any sesquilinear (bilinear) continuous form $c(x, y)$ defined on a pair of Hilbert spaces X, Y. Let

$$\inf_{x \in X} \sup_{y \in Y} \frac{|c(x, y)|}{\|x\|_X \|y\|_Y} > 0.$$

Then

$$\inf_{x \in X} \sup_{y \in Y} \frac{|c(x, y)|}{\|x\|_X \|y\|_Y} = \inf_{[y] \in Y/Y_0} \sup_{x \in X} \frac{|c(x, y)|}{\|x\|_X \|[y]\|_{Y/Y_0}}, \qquad (4.3.17)$$

where

$$Y_0 := \{y \in Y : c(x, y) = 0 \quad \forall x \in X\}$$

---

[15]In particular, $V_0 := \{v \in V : b(v, q) = 0 \quad \forall q \in Q\}$.

and Y/Y_0 is the quotient space whose elements are equivalence classes,

$$[y] = y + Y_0, \qquad \|[y]\|_{Y/Y_0} := \inf_{z \in [y]} \|z\|_Y .$$

The property is a direct consequence of the Banach Closed Range Theorem for continuous operators; compare Exercise 4.3.1.

We shall discuss now how the assumptions of the Babuška–Nečas and Babuška Theorems translate into appropriate assumptions on forms $a(u,v), b(u,q)$. We shall assume that a "big" sesquilinear form satisfies the inf-sup condition,

$$\inf_{\mathsf{u}} \sup_{\mathsf{v}} \frac{|\mathsf{b}(\mathsf{u},\mathsf{v})|}{\|\mathsf{u}\| \, \|\mathsf{v}\|} =: \gamma > 0 . \qquad (4.3.18)$$

Setting $\mathsf{u} = (0,p)$ in (4.3.18), we get

$$\sup_{(v,q)} \frac{|b^*(p,v)|}{(\|v\|^2 + \|q\|^2)^{1/2}} = \sup_{v} \frac{|b^*(p,v)|}{\|v\|} = \sup_{v} \frac{|b(v,p)|}{\|v\|} \geq \gamma \|p\| .$$

The condition

$$\sup_{v} \frac{|b(v,p)|}{\|v\|} \geq \beta \|p\|, \quad p \in Q, \quad \beta > 0 , \qquad (4.3.19)$$

is the famous **BB** (Babuška–Brezzi)[16] or *the* inf-sup condition relating spaces V and Q. Note that $\beta \geq \gamma$.

The inf-sup condition for form $b(\mathsf{u},\mathsf{v})$ implies uniqueness,

$$b(\mathsf{u},\mathsf{v}) = 0 \quad \forall \mathsf{v} \qquad \Rightarrow \qquad \mathsf{u} = 0 .$$

Applying the statement to $\mathsf{u} = (u_0,p)$ where $u_0 \in V_0$, we obtain

$$a(u_0,v) + b^*(p,v) = 0 \quad \forall v \in V \qquad \Rightarrow \qquad u_0 = 0 \text{ and } p = 0 . \qquad (4.3.20)$$

Assume now that

$$a(u_0,v_0) = 0 \quad \forall v_0 \in V_0 . \qquad (4.3.21)$$

The BB condition (4.3.19) now implies that there exists a unique $p \in Q$ such that

$$\begin{cases} p \in Q , \\ b^*(p,v) = -a(u_0,v), \quad v \in V . \end{cases}$$

Indeed, according to assumption (4.3.21), the right-hand side in the equation above satisfies the required compatibility condition. The pair (u_0,p) satisfies thus the assumption in the uniqueness condition (4.3.20) and, therefore, $u_0 = 0$. In other words, we have the *uniqueness in kernel condition*,

$$a(u_0,v_0) = 0 \quad \forall v_0 \in V_0 \qquad \Rightarrow \qquad u_0 = 0 , \qquad (4.3.22)$$

i.e., operator

$$A_0 : V_0 \to V_0', \quad \langle A_0 u_0, v_0 \rangle = a(u_0,v_0), \quad u_0, v_0 \in V_0 ,$$

is injective.

Next, restricting ourselves in (4.3.18) to $\mathsf{u} = (u_0,p)$, $u_0 \in V_0$, we have

$$\sup_{(v,q)} \frac{|a(u_0,v) + b^*(p,v)|}{(\|v\|^2 + \|q\|^2)^{1/2}} = \sup_{v} \frac{|a(u_0,v) + b^*(p,v)|}{\|v\|} \geq \gamma (\|u_0\|^2 + \|p\|^2)^{1/2} .$$

---

[16]Sometimes also called the LBB (Ladyshenskaya–Babuška–Brezzi) condition.

Therefore,

$$\inf_{u_0 \in V_0, \, p \in Q} \sup_{v \in V} \frac{|a(u_0, v) + b^*(p, v)|}{(\|u_0\|^2 + \|p\|^2)^{1/2} \, \|v\|} = \inf_{[v] \in V/V_{00}} \sup_{u_0 \in V_0, \, p \in Q} \frac{|a(u_0, v) + b^*(p, v)|}{(\|u_0\|^2 + \|p\|^2)^{1/2} \, \|[v]\|} \geq \gamma,$$

where

$$V_{00} = \{v \in V \, : \, a(u_0, v) + b^*(p, v) = 0 \quad \forall u_0 \in V_0, \forall p \in Q\}$$

$$= \{v_0 \in V_0 \, : \, a(u_0, v_0) = 0 \quad \forall u_0 \in V_0\}.$$

This implies (the infimum is now taken over a smaller set) that

$$\inf_{[v_0] \in V_0/V_{00}} \sup_{u_0 \in V_0, \, p \in Q} \frac{|a(u_0, v_0)|}{(\|u_0\|^2 + \|p\|^2)^{1/2} \, \|[v_0]\|} = \inf_{[v_0] \in V_0/V_{00}} \sup_{u_0 \in V_0} \frac{|a(u_0, v_0)|}{\|u_0\| \, \|[v_0]\|} \geq \gamma.$$

Finally, uniqueness in kernel (4.3.22) implies that

$$\inf_{[v_0] \in V_0/V_{00}} \sup_{u_0 \in V_0} \frac{|a(u_0, v_0)|}{\|u_0\| \, \|[v_0]\|_{V_0/V_{00}}} = \inf_{u_0 \in V_0} \sup_{[v_0] \in V_0/V_{00}} \frac{|a(u_0, v_0)|}{\|u_0\| \, \|[v_0]\|}.$$

But (compare Exercise 4.3.2)

$$\sup_{v_0 \in V_0} \frac{|a(u_0, v_0)|}{\|v_0\|} = \sup_{[v_0] \in V_0/V_{00}} \frac{|a(u_0, v_0)|}{\|[v_0]\|}.$$

Consequently,

$$\inf_{u_0 \in V_0} \sup_{v_0 \in V_0} \frac{|a(u_0, v_0)|}{\|u_0\| \, \|v_0\|} = \inf_{[v_0] \in V_0/V_{00}} \sup_{u_0 \in V_0} \frac{|a(u_0, v_0)|}{\|u_0\| \, \|[v_0]\|_{V_0/V_{00}}} \geq \gamma,$$

i.e., the *inf-sup in kernel condition* holds:

$$\sup_{v_0 \in V_0} \frac{|a(u_0, v_0)|}{\|v_0\|} \geq \alpha \|u_0\|, \quad u_0 \in V_0, \tag{4.3.23}$$

with $\alpha \geq \gamma$.

On the discrete level, uniqueness implies existence for any right-hand side. Note that on the continuous level the null space of the transpose operator is

$$\{v \, : \, b(u, v) = 0 \quad \forall u\}$$

$$= \{(v, q) \in V \times Q \, : \, a(u, v) + b^*(p, v) + b(u, q) = 0 \quad \forall u \in V, \forall p \in Q\}$$

$$= \{(v_0, q) \in V_0 \times Q \, : \, a(u, v_0) + b(u, q) = 0 \quad \forall u \in V\}$$

$$= \{(v_0, q_0) \in V_{00} \times Q\},$$

where, in the last line, $q_0 \in Q$ is the unique solution of the problem

$$\begin{cases} q_0 \in Q, \\ b(u, q_0) = -a(u, v_0) \quad \forall u \in V, \quad v_0 \in V_{00}. \end{cases}$$

In order for the mixed problem to have a solution, the right-hand side must satisfy the compatibility condition:

$$f(v_0) + g(q_0) = 0 \quad \forall v_0 \in V_{00}. \tag{4.3.24}$$

Brezzi $\Rightarrow$ Babuška. Assume now that Brezzi's conditions (4.3.23) and (4.3.19) hold. We shall demonstrate now that the two "small" inf-sup conditions imply that the "big" condition (4.3.18) must be satisfied as well. Given $(u, p) \in V \times Q$, define

$$f(v) := a(u, v) + b^*(p, v), \quad v \in V,$$
$$g(q) := b(u, q), \qquad\qquad q \in Q.$$

We need to demonstrate that we control $\|u\|$ and $\|p\|$ by norms of f, g.

The BB condition implies that

$$\inf_{[v]} \sup_{q} \frac{|b(v, q)|}{\|[v]\|_{V/V_0} \|q\|} = \inf_{q} \sup_{v} \frac{|b(v, q)|}{\|v\| \|q\|} = \beta > 0.$$

Consequently,

$$\|[u]\|_{V/V_0} = \inf_{w \in V_0} \|u - w\|_V = \|u - u_0\| \leq \frac{1}{\beta} \|g\|_{Q'}$$

where u_0 is the V-orthogonal projection of u onto V_0. Now,

$$a(u, v_0) = a(u - u_0, v_0) + a(u_0, v_0) = f(v_0), \quad v_0 \in V_0,$$

and the inf-sup in kernel condition gives

$$\|u_0\| \leq \frac{1}{\alpha} (\|a\| \|u - u_0\| + \|f\|_{V_0'})$$
$$\leq \frac{1}{\alpha} \left(\frac{\|a\|}{\beta} \|g\|_{Q'} + \|f\|_{V'} \right).$$

Consequently,

$$\|u\| \leq \|u_0\| + \|u - u_0\| \leq \frac{1}{\alpha} \|f\|_{V'} + \frac{1}{\beta} \left(1 + \frac{\|a\|}{\alpha} \right) \|g\|_{Q'}. \tag{4.3.25}$$

Finally, we can use the first equation and the BB condition to control the Lagrange multiplier p.

$$b^*(p, v) = f(v) - a(u, v)$$

implies

$$\|p\| \leq \frac{1}{\beta} (\|f\|_{V'} + \|a\| \|u\|)$$
$$\leq \frac{1}{\beta} \left(1 + \frac{\|a\|}{\alpha} \right) \|f\|_{V'} + \frac{\|a\|}{\beta^2} \left(1 + \frac{\|a\|}{\alpha} \right) \|g\|_{Q'}. \tag{4.3.26}$$

We can formulate now the famous Brezzi Theorem.

Theorem 4.12 (Brezzi [11]). *Assume that Brezzi's conditions (4.3.23) and (4.3.19) hold on both continuous and discrete levels and that discrete inf-sup constants remain uniformly bounded away from zero,*

$$\beta_h \geq \beta_0 > 0, \quad \alpha_h \geq \alpha_0 > 0.$$

Let $f \in V'$, $g \in Q'$ satisfy the compatibility condition (4.3.24). Then both continuous and discrete problems are well-posed, i.e., there exist unique solutions (u, p) and (u_h, p_h) and stability constants $\gamma = \gamma(\alpha, \beta, \|a\|)$ and $\gamma_0 = \gamma(\alpha_0, \beta_0, \|a\|)$ such that

$$\|(u, p)\| \leq \frac{1}{\gamma} (\|f\|_{V'}^2 + \|g\|_{Q'}^2)^{1/2}$$

and

$$\|(u_h, p_h)\| \leq \frac{1}{\gamma_0} (\|f\|_{V'}^2 + \|g\|_{Q'}^2)^{1/2}.$$

Moreover, the following error estimate holds:

$$\left(\|v - v_h\|_V^2 + \|p - p_h\|_Q^2 \right) \leq \frac{\|b\|}{\gamma_0} \left(\inf_{w_h \in V_h} \|v - w_h\|_V^2 + \inf_{p_h \in Q_h} \|p - p_h\|_Q^2 \right). \tag{4.3.27}$$

Remark 4.13. Continuity constant $\|b\|$ can be easily bounded by the continuity constants for the small forms, e.g., $\|b\| \leq \|a\| + 2\|b\|$. We have shown that Brezzi's conditions *are not only sufficient but also necessary* for the well-posedness of the mixed problem, discrete stability, and convergence.

As usual, the inf-sup condition on the continuous level does not imply the corresponding discrete inf-sup condition. However, there is a general tool that helps to relate the two conditions.

4.3.1 ▪ Fortin Operator

Let $b(v, q)$ be a bilinear or sesquilinear form defined on a pair of Hilbert spaces V, Q. Assume that $b(v, p)$ satisfies the inf-sup condition,

$$\sup_{v \in V} \frac{|b(v, p)|}{\|v\|_V} \geq \beta \|p\|_Q, \quad p \in Q, \quad \beta > 0. \tag{4.3.28}$$

Let $V_h \subset V, Q_h \subset Q$ be a pair of discrete spaces. A linear and continuous operator

$$\Pi_h : V \to V_h, \quad \|\Pi_h v\|_V \leq \|\Pi_h\| \|v\|_V,$$

is called a *Fortin operator* if the following discrete orthogonality condition holds:

$$b(v - \Pi_h v, p_h) = 0, \quad v \in V, \, p_h \in Q_h. \tag{4.3.29}$$

We then have

$$\begin{aligned}
\sup_{v_h \in V_h} \frac{|b(v_h, p_h)|}{\|v_h\|_V} &\geq \sup_{v \in V} \frac{|b(\Pi_h v, p_h)|}{\|\Pi_h v\|_V} \\
&= \sup_{v \in V} \frac{|b(v, p_h)|}{\|v\|_V} \frac{\|v\|_V}{\|\Pi_h v\|_V} \\
&\geq \frac{\beta}{\|\Pi_h\|} \|p_h\|_Q.
\end{aligned}$$

In other words, the discrete inf-sup condition holds with a discrete inf-sup constant $\beta_h \geq \beta/\|\Pi_h\|$.

The concept of Fortin operator provides a general framework for proving discrete stability but a concrete construction of such an operator is problem dependent. Note that the Fortin operator needs to be defined on the *whole* energy space and, therefore, one cannot use standard interpolation operators that are defined typically only for sufficiently regular functions.

We shall show now an example of such a construction for the Stokes problem.

4.3.2 ▪ Example of a Stable Pair for the Stokes Problem

We can give now perhaps the simplest example of a stable pair of elements for the Stokes problem; see Section 1.4.2. Let $\Omega \subset \mathbb{R}^2$ be a polygonal domain with a boundary split into nonzero measure parts of Γ_u and Γ_t. To fit the problem into the Brezzi theory, define

$$V := \{u \in (H^1(\Omega))^2 : u = 0 \text{ on } \Gamma_u\},$$

$$Q := L^2(\Omega),$$

$$a(u, v) := \mu \int_\Omega (\boldsymbol{\nabla} u + \boldsymbol{\nabla}^T u) : \boldsymbol{\nabla} v, \qquad u, v \in V, \quad \mu > 0,$$

$$b(u, q) := \int_\Omega \operatorname{div} u \, q, \quad u \in V, q \in Q,$$

$$f(v) := \int_\Omega fv + \int_{\Gamma_t} tv,$$

$$g(v) := \int_\Omega gv,$$

where $f, g \in L^2(\Omega)$, $t \in L^2(\Gamma_t)$ are given. All spaces are real.

Notice that

$$a(u, v) = 2\mu \int_\Omega \epsilon_{ij}(u)\epsilon_{ij}(v).$$

The essential BC on Γ_u and the Korn inequality imply thus that form $a(u, v)$ is coercive on the whole space V. The *inf-sup in kernel condition* is thus trivially satisfied. The LBB condition is a subject of a major theorem.

Theorem 4.14. *Let $\Omega \subset \mathbb{R}^N$ be a Lipschitz domain with boundary Γ split into parts Γ_1 and Γ_2. There exists then a constant $\beta > 0$ such that the following hold:*

- Case: $\operatorname{meas}(\Gamma_2) > 0$,

$$\sup_{\substack{v \in H^1(\Omega)^N \\ v = 0 \text{ on } \Gamma_1}} \frac{|\int_\Omega p \operatorname{div} v|}{\|v\|_{H^1(\Omega)}} \ge \beta\|p\|_{L^2(\Omega)}, \qquad p \in L^2(\Omega). \tag{4.3.30}$$

- Case: $\Gamma_1 = \Gamma$,

$$\sup_{v \in H^1_0(\Omega)^N} \frac{|\int_\Omega p \operatorname{div} v|}{\|v\|_{H^1(\Omega)}} \ge \beta\|p\|_{L^2(\Omega)}, \qquad p \in L^2_0(\Omega). \tag{4.3.31}$$

The proof is very nontrivial and we will not provide it; see, e.g., [64]. See also [21] for connections with other inequalities and historical comments. The continuous problem is thus well-posed.

We discretize the velocities with quadratic triangular Lagrange elements, and the pressure with piecewise constants defined on the same mesh. We will use now the Fortin "trick" to demonstrate that the pair satisfies the discrete inf-sup condition. Define a candidate for the Fortin operator,

$$\Pi_h v := \Pi_1 v + \Pi_2(v - \Pi_1 v), \qquad v \in H^1(\Omega), \tag{4.3.32}$$

where Π_1 is the modified Clément interpolation operator presented in Section 3.6. The "correcting" operator,

$$\Pi_2 : (H^1(K))^2 \to (\mathcal{P}^2(K))^2,$$

is a local, linear operator defined by requesting two conditions:

$$\Pi_2 v = 0 \quad \text{at vertices} \quad \text{and} \quad \int_e (v - \Pi_2 v) = 0 \quad \text{for each edge } e\,.$$

Note that

$$v - \Pi_h v = (v - \Pi_1 v) - \Pi_2(v - \Pi_1 v) = (I - \Pi_2)(v - \Pi_1 v)\,.$$

The modified Clément operator preserves the homogeneous Dirichlet BC, i.e., if $v = 0$ on a boundary edge e, then $\Pi_1 v = 0$ on e as well. Linearity of Π_2 implies that $\Pi_h v = 0$ vanishes then, too.

It follows from the construction of Π_2 that, for a constant q,

$$\int_K \operatorname{div}(v - \Pi_h v)\,q = q \sum_e \int_e (v - \Pi_h v) \cdot n = q \sum_e \int_e (I - \Pi_2)(v - \Pi_1 v) \cdot n = 0\,.$$

At the same time, a standard scaling argument implies that

$$\|\Pi_2 v\|_{L^2(K)}^2 + |\Pi_2 v|_{H^1(K)}^2 \lesssim h_K^2 \|\widehat{\Pi_2 v}\|_{L^2(\hat{K})}^2 + |\widehat{\Pi_2 v}|_{H^1(\hat{K})}^2$$

$$= h_K^2 \|\widehat{\Pi}_2 \hat{v}\|_{L^2(\hat{K})}^2 + |\widehat{\Pi}_2 \hat{v}|_{H^1(\hat{K})}^2$$

$$\lesssim \|\hat{v}\|_{H^1(\hat{K})}^2$$

$$\lesssim C(h_K^{-2}\|v\|_{L^2(K)}^2 + |v|_{H^1(K)}^2)\,,$$

where, as usual, $A \lesssim B$ means existence of a constant C (independent of element and function v) such that $A \le CB$. The estimate above, combined with estimate (3.6.82), proves the continuity of operator Π_h.

4.3.3 ▪ Time-Harmonic Maxwell Equations as an Example of a Mixed Problem

Another example of a mixed problem is provided by the Maxwell equations. Recall the stabilized variational formulation for time-harmonic Maxwell equations (1.4.64),[17]

$$\begin{cases} E \in H_0(\operatorname{curl}, \Omega),\ p \in H_0^1(\Omega)\,, \\[2mm] \displaystyle\int_\Omega \frac{1}{\mu}\nabla \times E \cdot \nabla \times \overline{F} - \omega^2 \int_\Omega \epsilon E \cdot \overline{F} + \int_\Omega \epsilon \nabla p \cdot \overline{F} = -i\omega \int_\Omega J^{\mathrm{imp}} \cdot \overline{F}\,, \\[3mm] \hspace{6cm} F \in H_0(\operatorname{curl}, \Omega)\,, \\[3mm] \displaystyle\int_\Omega \epsilon E \cdot \nabla \overline{q} = 0, \qquad q \in H_0^1(\Omega)\,, \end{cases} \qquad (4.3.33)$$

where $\operatorname{div} J^{\mathrm{imp}} = 0$ and

$$0 < \mu_0 \le \mu \le \mu_\infty < \infty, \qquad 0 < \epsilon_0 \le \epsilon \le \epsilon_\infty < \infty\,.$$

Using the notation for the mixed problems, we have

$$a(E, F) := \int_\Omega \frac{1}{\mu}\nabla \times E \cdot \nabla \times \overline{F} - \omega^2 \int_\Omega \epsilon E \cdot \overline{F}\,,$$

$$b(E, q) := \int_\Omega \epsilon E \cdot \nabla \overline{q}\,,$$

$$l(F) := -i\omega \int_\Omega J^{\mathrm{imp}} \cdot \nabla \overline{F}\,.$$

---

[17]Case $\Gamma_H = \emptyset$.

Compared with the Stokes problem, the difficulties are now completely reversed. For Stokes, form $a(u, v)$ was V-coercive (so the inf-sup in kernel condition was trivially satisfied) but proving the BB condition was a challenge. For Maxwell, the BB condition is simple as it is a direct consequence of the exact sequence property. Indeed, with the homogeneous BCs, the standard norm in $H_0^1(\Omega)$ is equivalent to the H^1-seminorm,

$$\|q\|_{H^1(\Omega)}^2 \sim \int_\Omega |\boldsymbol{\nabla} q|^2 .$$

We now have

$$\sup_F \frac{|\int_\Omega \epsilon \boldsymbol{\nabla} p \cdot F|}{\|F\|_{H(\mathrm{curl},\Omega)}} \geq \frac{|\int_\Omega \epsilon \boldsymbol{\nabla} p \cdot \boldsymbol{\nabla} p|}{\|\boldsymbol{\nabla} p\|_{L^2(\Omega)}} \geq \epsilon_0 \|\boldsymbol{\nabla} p\|_{L^2} \sim \epsilon_0 \|p\|_{H^1(\Omega)}$$

since $\boldsymbol{\nabla} H_0^1(\Omega) \subset H(\mathrm{curl}, \Omega)$. Note that the same reasoning applies to the discrete problem discussed below.

On the other side, proving the discrete inf-sup in kernel condition is a challenge. The kernel

$$Q_0 := \{E \in H_0(\mathrm{curl}, \Omega) \: : \: (\epsilon E, \boldsymbol{\nabla} q) = 0, \quad q \in H_0^1(\Omega)\}$$

consists of all fields in $H_0(\mathrm{curl}, \Omega)$ with vanishing divergence, $\mathrm{div}(\epsilon E) = 0$. One way to analyze the inf-sup in kernel condition is to introduce (generalized) eigenvalue problems:

$$\begin{cases} e_i \in Q_0, \, \lambda_i \in \mathbb{C}, \\ (\mu^{-1} \boldsymbol{\nabla} \times e_i, \boldsymbol{\nabla} \times F) = \lambda_i (\epsilon \, e_i, F), \quad F \in Q_0 . \end{cases} \tag{4.3.34}$$

The self-adjointness and positive semidefiniteness of the curl-curl operator imply that the eigenvalues are real and nonnegative. Under the additional assumption that Ω is simply connected, one can show that all eigenvalues are positive and form a sequence converging to infinity,

$$\lambda_i = \omega_i^2, \quad 0 < \omega_i \to \infty \text{ as } i \to \infty,$$

with corresponding finite-dimensional eigenspaces. Without losing generality, we can assume the existence of eigenpairs (e_i, ω_i^2) such that e_i provide an orthonormal (in terms of both forms) basis for V_0. Introducing spectral components for E and F,

$$E = \sum_{i=1}^\infty E_i e_i, \quad F = \sum_{j=1}^\infty F_j e_j,$$

we end up with the spectral representation,

$$a(E, F) = \sum_{i=1}^\infty (\omega_i^2 - \omega^2) E_i \overline{F}_i .$$

Now, over the kernel, the curl-curl term represents a norm equivalent to the $H(\mathrm{curl})$-norm. Indeed,

$$\left(\frac{1}{\mu} \boldsymbol{\nabla} \times E, \boldsymbol{\nabla} \times E \right) \leq \frac{1}{\mu_0} \|\boldsymbol{\nabla} \times E\|^2 \leq \frac{1}{\mu_0} \|E\|_{H(\mathrm{curl},\Omega)}^2 .$$

At the same time,

$$
\|\nabla \times E\|^2 + \|E\|^2 \leq \min\left\{\frac{1}{\mu_\infty}, \epsilon_0\right\}\left[\left(\frac{1}{\mu}\nabla \times E, \nabla \times E\right) + (\epsilon E, E)\right]
$$

$$
\leq \min\left\{\frac{1}{\mu_\infty}, \epsilon_0\right\} \sum_{i=1}^{\infty}(\omega_i^2 + 1)|E_i|^2
$$

$$
\leq \min\left\{\frac{1}{\mu_\infty}, \epsilon_0\right\} \frac{2}{\omega_1^2} \sum_{i=1}^{\infty} \omega_i^2 |E_i|^2
$$

$$
= \min\left\{\frac{1}{\mu_\infty}, \epsilon_0\right\} \frac{2}{\omega_1^2}\left(\frac{1}{\mu}\nabla \times E, \nabla \times E\right).
$$

Replacing the $H(\mathrm{curl})$-norm with $(\sum_{i=1}^{\infty}\omega_i^2|E_i|^2)^{1/2}$, we can compute now explicitly the inf-sup in kernel constant exactly in the same way as for the model vibration problem in Section 4.2, obtaining

$$
\alpha = \min_i \frac{|\omega_i^2 - \omega^2|}{\omega_i^2}.
$$

This finishes the proof that the continuous mixed problem is well-posed. We proceed now with the discretization, introducing discrete subspaces

$$
W_h \subset H_0^1(\Omega) \quad \text{and} \quad Q_h \subset H_0(\mathrm{curl}, \Omega)
$$

to arrive at the discretization of (4.3.33),

$$
\begin{cases}
E_h \in Q_h, \, p_h \in W_h, \\
a(E_h, F_h) + b^*(p_h, F_h) = l(F_h), \quad F_h \in Q_h, \\
\quad b(E_h, q_h) = 0, \quad q_h \in W_h.
\end{cases} \tag{4.3.35}
$$

Note that in choosing the notation, we are following now the notation for the exact sequence spaces rather than the abstract mixed problem. As mentioned above, the discrete version of the BB inf-sup condition is a consequence of the exact sequence property: $\nabla W_h \subset Q_h$. To investigate the discrete version of the inf-sup in kernel condition, we can repeat the same reasoning as on the continuous level. We begin by introducing the discrete kernel,

$$
Q_{h,0} := \{E_h \in Q_h \,:\, b(E_h, q_h) = 0, \quad q_h \in W_h\},
$$

and the corresponding Galerkin discretization of eigenvalue problem (4.3.34),

$$
\begin{cases}
e_{i,h} \in Q_{h,0}, \, \omega_{i,h} \in \mathbb{R}_+, \\
(\mu^{-1}\nabla \times e_{i,h}, \nabla \times F_h) = \omega_{i,h}^2(\epsilon \, e_{i,h}, F_h), \quad F_h \in Q_{h,0}.
\end{cases}
$$

Repeating the reasoning from the continuous level, we obtain the analogous formula for the discrete inf-sup constant,

$$
\alpha_h = \min_i \frac{\omega_{i,h}^2 - \omega^2}{\omega_i^2}.
$$

If the discrete eigenvalues converge to the exact ones, the discrete inf-sup constant converges to the exact one. As for the Helmholtz equation, the discrete stability has clearly an asymptotic character and we can also recall the criterion for reaching the asymptotic stability: *all discrete eigenvalues must be on the same side of ω as the exact eigenvalues*. This is, unfortunately, where the analogy with the Helmholtz problem stops. The discrete Maxwell eigenvalue problem is solved in space $Q_{h,0}$ which is *not* a subspace of the continuous kernel Q_0. The proof of convergence of discrete eigenvalues to the continuous ones is much more difficult than for elliptic problems and it involves a concept of *discrete compactness*; see [59, 8] and the literature therein. Contrary to elliptic problems, convergence of discrete eigenvalues to the exact ones *may not be monotone*; see experiments in [38].

Exercises

4.3.1. Use the Banach Closed Range Theorem to prove (4.3.17). (3 points)

4.3.2. Let $f \in X'$ be an (anti)linear and continuous functional defined on a normed space X. Let $M \subset \mathcal{N}(f)$ be a subspace of the null space of f. Define

$$\hat{f} : X/M \to \mathbb{R}(\mathbb{C}), \quad \langle \hat{f}, [x] \rangle := \langle f, y \rangle, \, y \in [x].$$

Explain why functional $\hat{f}$ is well-defined and prove that

$$\sup_{x \in X} \frac{|\langle f, x \rangle|}{\|x\|_X} =: \|f\|_{X'} = \|\hat{f}\|_{(X/M)'} := \sup_{[x] \in X/M} \frac{|\langle \hat{f}, [x] \rangle|}{\|[x]\|_{X/M}}.$$

(2 points)

4.3.3. In the Hilbert space setting, it is elegant to preserve the Hilbert space structure by introducing the Euclidean norm for the group variable:

$$\|\mathsf{u}\|^2 = \|(u,p)\|^2 := \|u\|_V^2 + \|p\|_Q^2.$$

Revisit the reasoning in the text in an attempt to derive sharper bounds for Babuška's inf-sup constant γ for form $b(\mathsf{u}, \mathsf{v})$ in terms of Brezzi's inf-sup constants α, β (and continuity constant $\|a\|$) using the Euclidean norms for u, v. *Hint:* By the Pythagoras Theorem we have

$$\|u\|_V^2 = \|u_0\|_V^2 + \|u - u_0\|_V^2.$$

(15 points)

4.3.4. Stokes problem with pure kinematic BCs. Consider the Stokes problem with kinematic homogeneous BC implied on the whole boundary, i.e.,

$$V = (H_0^1(\Omega))^N, \quad N = 2, 3.$$

Note that the pressure cannot now be unique as, for any constant p,

$$\int_\Omega p \operatorname{div} v = p \int_\Gamma v \cdot n = 0 \quad \forall v \in V.$$

This leads to the modification of space Q; the L^2-space is replaced with the quotient space $L^2(\Omega)/\mathbb{R}$ that is isomorphic and isometric with the subspace of $L^2(\Omega)$ consisting of functions of zero average,

$$Q = L^2(\Omega)/\mathbb{R} \sim L_0^2(\Omega) := \left\{ q \in L^2(\Omega) : \int_\Omega q = 0 \right\}.$$

Use Brezzi's Theorem to prove that the problem is well-posed. Does the Fortin operator discussed in the text work for this problem as well? (2 points)

4.3.5. Let I be a unit interval covered with a uniform FE mesh of N elements of order $p = 1$ or $p = 2$. Let V_h denote the corresponding FE space of H^1-conforming elements. Prove the following upper bound:

$$\|v_h\|_{H^{1/2}(I)} \lesssim h^{-1/2}\|v_h\|_{L^2(I)}, \quad v_h \in V_h.$$

Let U_h be now the FE space spanned by piecewise constants defined on the same mesh. Use the duality argument to prove the lower bound,

$$h^{1/2}\|u_h\|_{L^2(I)} \lesssim \|u_h\|_{\tilde{H}^{-1/2}(I)}, \quad u_h \in U_h.$$

(5 points)

4.3.6. Unstable Petrov–Galerkin discretization of the $H^{1/2}$-$\tilde{H}^{-1/2}$ duality pairing. Consider a unit interval $I = (0,1)$ discretized with a uniform mesh of N elements. Introduce two discrete spaces: test space V_h spanned by the standard $N + 1$ "hat functions," and trial space U_h spanned by N piecewise constants defined on the same mesh. Prove that the discrete inf-sup constant γ_h in the inf-sup condition

$$\sup_{v_h \in V_h} \frac{|\int_I u_h v_h \, dx|}{\|v_h\|_{H^{1/2}(I)}} \geq \gamma_h \|u_h\|_{\tilde{H}^{-1/2}(I)}$$

is of order $h^{1/2-\epsilon}$ and, therefore, the corresponding discretization of the duality pairing is unstable. *Hint:* Use the oscillating trial function taking ± 1 values to derive the upper bound for γ_h. (15 points)

4.3.7. Stable Petrov–Galerkin discretization of the $H^{1/2}$-$\tilde{H}^{-1/2}$ duality pairing. Consider a unit interval $I = (0,1)$ discretized with a uniform mesh of N elements. Introduce two discrete spaces: test space V_h spanned by the standard $N + 1$ "hat functions," and trial space U_h spanned by $N + 1$ piecewise constants defined on the *dual mesh*; see Fig. 4.3. Prove the discrete inf-sup condition

$$\sup_{v_h \in V_h} \frac{|\int_I u_h v_h \, dx|}{\|v_h\|_{H^{1/2}(I)}} \geq \gamma \|u_h\|_{\tilde{H}^{-1/2}(I)}$$

with mesh-independent constant $\gamma > 0$. *Hint:* You may need the Gershgorin's Theorem (see, e.g., [54, p. 287]). (15 points)

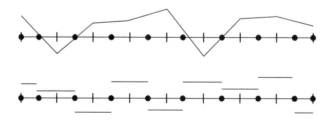

Figure 4.3: Test and trial discrete spaces defined on primal and dual meshes.

4.4 ▪ Nonuniform Meshes

This section deals with a coercive problem and as such belongs to Chapter 2 on coercive problems. The reason I have put it into this chapter is twofold: (a) the theory uses *weighted Sobolev spaces* which we have not discussed yet, and (b) it is a more advanced subject that I teach only occasionally. The presented theory is about the estimation of the interpolation error for functions with singularities and applies to noncoercive problems experiencing such solutions, as well.

As we have learned, in the presence of singularities, the rate of convergence for uniform h-refinements is limited not by the polynomial degree but rather by the global regularity of the solution expressed in terms of Sobolev norms. In this section, we will present the fundamental result of Babuška, Kellogg, and Pitkäranta [5] demonstrating that, by using properly designed nonuniform meshes, graded toward the singular points, one can restore the optimal rate of convergence dictated by the polynomial degree p alone. In practice, the meshes are obtained by using a posteriori error estimates and automatic h-adaptivity. The proof will deal with a simple 2D model problem, triangular elements, and polynomial order $p = 1$ only, but the result has been numerically confirmed for elements of all shapes, and 3D elliptic problems [25, 35].

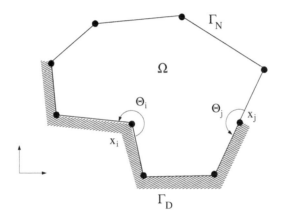

Figure 4.4: A 2D polygonal domain.

Let Ω be a 2D polygonal domain (see Fig. 4.4) with vertices x_i and corresponding internal angles θ_i, $i = 1, \ldots, M$. Let boundary Γ be partitioned into a Dirichlet boundary Γ_D and Neumann boundary Γ_N. Note that Γ_D may be terminated inside of an edge, in which case the endpoint of Γ_D is classified also as a vertex; see vertex x_j in Fig. 4.4. For each vertex x_i introduce a parameter α_i,

$$\alpha_i := \min\left\{1, \frac{\kappa_i \pi}{\theta_i}\right\},$$

where $\kappa_i = 1$ if both sides of x_i are contained either in Γ_D or Γ_N, and $\kappa_i = 1/2$ if vertex x_i is a transition point between the two parts of the boundary. Let $\mathcal{M}$ denote the subset of ("singular") vertices for which coefficients $\alpha_i < 1$. Let $\beta := (\beta_1, \ldots, \beta_M)$ be an M-tuple of exponents associated with the vertices, $\beta_i \in [0, 1)$. Consider the weight function

$$\phi_\beta(x) := \prod_{i=1}^{M} |x - x_i|^{\beta_i}, \qquad (4.4.36)$$

where $|x|$ denotes the Euclidean norm of vector x. Let $H^m(\Omega), m = 1, 2, \ldots,$ denote the regular Sobolev spaces, and

$$H_D^1(\Omega) := \{u \in H^1(\Omega) : u = 0 \text{ on } \Gamma_D\}.$$

We will consider the model problem

$$\begin{cases} u \in H_D^1(\Omega)\,, \\ \displaystyle\int_\Omega \boldsymbol{\nabla} u \cdot \boldsymbol{\nabla} v + uv = \int_\Omega fv\,, \qquad v \in H_D^1(\Omega)\,. \end{cases} \qquad (4.4.37)$$

In the presence of singularities, regularity of the solution can be assessed using the *weighted Sobolev norms*:

$$\|u\|_{H^{m,\beta}(\Omega)}^2 := \|u\|_{H^{m-1}(\Omega)}^2 + \underbrace{\int_\Omega \phi_\beta^2 \sum_{|\alpha|=m} |D^\alpha u|^2}_{=: |u|_{H^{m,\beta}(\Omega)}^2}\,.$$

For $\beta = 0$, the norm coincides with the standard Sobolev norm. *Completion* of $C^\infty(\bar\Omega)$ under the weighted norm is identified as the *weighted Sobolev space* and denoted by $H^{m,\beta}(\Omega)$. Note that the weight applies only to the highest order derivatives. One can prove the continuous embedding:

$$H^{m,\beta}(\Omega) \hookrightarrow C^{m-2}(\bar\Omega)\,, \quad m \geq 2\,.$$

The following regularity result has been established in [5].

Theorem 4.15. *Assume that*

$$1 - \alpha_i < \beta_i < 1\,, \quad x_i \in \mathcal{M}\,,$$

and $f \in H^{0,\beta}(\Omega)$. The solution u lives then in $H_D^1(\Omega) \cap H^{2,\beta}(\Omega)$, and

$$\|u\|_{H^{2,\beta}(\Omega)} \leq C \|f\|_{H^{0,\beta}(\Omega)}$$

with stability constant C independent of f.

Note that, for each nonsingular vertex x_i, we can select $\beta_i = 0$. The considered model problem is (trivially) $H^1(\Omega)$-coercive and the convergence analysis reduces to the interpolation error estimates. Following [5], we will consider a special class of nonuniform meshes whose density is controlled by a weight function ϕ_γ. Hereafter γ will denote a generic M-tuple; $\gamma = \beta$ for the problem of interest.

Definition. Let $h, L > 0$. Triangulation $\mathcal{T}$ is of type (h, γ, L) if the following three conditions are satisfied:

(i) Minimum angle condition:

$$\theta \geq L^{-1} \quad \forall \text{ angle } \theta \text{ of } T, \quad \forall \text{ element } T \in \mathcal{T}\,.$$

(ii) Control of element size for elements with positive weight. If $\phi_\gamma \neq 0$ on $\bar{T}$, then

$$L^{-1} h \sup_{x \in T} \phi_\gamma(x) \leq d_T \leq Lh \inf_{x \in T} \phi_\gamma(x)\,.$$

(iii) Control of element size for elements with vanishing weight. If $\phi_\gamma = 0$ at some point in $\bar{T}$, then

$$L^{-1} h \sup_{x \in T} \phi_\gamma(x) \leq d_T \leq Lh \sup_{x \in T} \phi_\gamma(x)\,.$$

Above, d_T denotes diameter of element T,

$$d_T := \sup_{x,y \in T} |x - y|\,.$$

Although several results discussed next will apply to a general γ, the final interpolation error estimate will be applied to $\gamma = \beta$, with $\beta_i > 0$ only at singular vertices $x_i \in \mathcal{M}$. Consequently, case (ii) above applies to elements that are *not* adjacent to a singular vertex, and case (iii) deals with elements sharing a singular vertex. For the domain illustrated in Fig. 4.4, we have only three "singular" vertices: x_i with a reentrant corner $\gamma_i > \pi$, and two transition points (including x_j) between Dirichlet and Neumann parts of the boundary.

Lemma 4.16. *The following inequalities hold:*

$$\int_0^1 s^{-2}|v(s)|^2\,ds \le 4\int_0^1 |v'(s)|^2\,ds\,,$$
$$v \in H^1(0,1),\ v(0) = 0\,,$$

$$\int_1^\infty s^{-2}|v(s)|^2\,ds \le \int_1^\infty (s-1)^{-2}|v(s)|^2\,ds \le 4\int_1^\infty |v'(s)|^2\,ds\,,$$
$$v \in H^1(1,\infty),\ v(1) = 0\,. \tag{4.4.38}$$

Notice in the first case that, by the 1D Poincaré inequality, the L^2-norm of v is bounded by the L^2-norm of v'. The estimate says that we control the stronger, weighted (with singular weight s^{-2}) L^2-norm of v as well.

Proof. The main tool in the proof is the *Integral Minkowski Inequality* (see [66, p. 409]),

$$\left(\int_a^b \left|\int_c^d f(t,s)\,ds\right|^2 dt\right)^{1/2} \le \int_c^d \left|\int_a^b f(t,s)\,dt\right|^{1/2} ds\,.$$

Representing $v(x)$ in terms of its derivative,

$$v(x) = \int_0^x v'(t)\,dt,$$

we have

$$\int_0^1 s^{-2}\left|\int_0^s v'(t)\,dt\right|^2 ds$$

$$= \int_0^1 \left|\int_0^1 v'(su)\,du\right|^2 ds \qquad \text{(change of variable: } t = su)$$

$$\le \left[\int_0^1 \left(\int_0^1 |v'(su)|^2\,ds\right)^{1/2} du\right]^2 \qquad \text{(Integral Minkowski Inequality)}$$

$$= \left[\int_0^1 \frac{1}{u^{1/2}}\left(\int_0^u |v'(t)|^2\,dt\right)^{1/2} du\right]^2 \qquad \text{(change of variables: } t = su)$$

$$\le \left[\int_0^1 \frac{1}{u^{1/2}}\left(\int_0^1 |v'(t)|^2\,dt\right)^{1/2} du\right]^2 \qquad \text{(upper bound)}$$

$$\le \int_0^1 |v'(t)|^2\,dt \left[\int_0^1 \frac{1}{u^{1/2}}\,du\right]^2$$

$$\le 4\int_0^1 |v'(t)|^2\,dt\,.$$

Similarly,

$$\int_1^\infty (s-1)^{-2} \left| \int_1^s v'(t)\,dt \right|^2 ds$$

$$= \int_1^\infty \left| \int_0^1 \frac{dv}{dt}(\xi(s-1)+1)\,d\xi \right|^2 ds \qquad \text{(change of variable: } t = \xi(s-1)+1\text{)}$$

$$\leq \left[\int_0^1 \left(\int_1^\infty \left| \frac{dv}{dt}(\xi(s-1)+1) \right|^2 ds \right)^{1/2} d\xi \right]^2 \qquad \text{(Integral Minkowski Inequality)}$$

$$= \left[\int_0^1 \xi^{-1/2} \left(\int_1^\infty \left| \frac{dv}{dt}(t) \right|^2 dt \right)^{1/2} d\xi \right]^2 \qquad \text{(change of variable: } t = \xi(s-1)+1\text{)}$$

$$= \int_1^\infty \left| \frac{dv}{dt}(t) \right|^2 dt \left(\int_0^1 \xi^{-1/2}\,d\xi \right)^2$$

$$= 4 \int_1^\infty \left| \frac{dv}{dt}(t) \right|^2 dt. \qquad \square$$

Lemma 4.17. *The following inequality holds:*

$$\int_0^1 t^{\alpha-2}[z(t)-a]^2\,dt \leq C(\alpha) \int_0^1 t^\alpha |z'(t)|^2\,dt, \qquad \alpha \neq 1, \qquad (4.4.39)$$

where

$$a = \begin{cases} z(0) & \text{for } \alpha < 1, \\ z(1) & \text{for } \alpha > 1. \end{cases}$$

Proof.
Case: $\alpha < 1$. Using the change of variables

$$t^{1-\alpha} = s, \quad s^{\frac{1}{1-\alpha}} = t, \quad (1-\alpha)t^{-\alpha}\,dt = ds,$$

we get

$$\int_0^1 t^{\alpha-2}[z(t)-z(0)]^2\,dt = \int_0^1 t^{2\alpha-2}[z(t)-z(0)]^2\,t^{-\alpha}\,dt$$

$$= (1-\alpha)^{-1} \int_0^1 s^{-2} \underbrace{[z(s^{\frac{1}{1-\alpha}})-z(0)]^2}_{=:v(s)}\,ds.$$

Using Lemma 4.16, we can bound the last integral by

$$4(1-\alpha)^{-1} \int_0^1 \left| \frac{dv}{ds} \right|^2 ds.$$

Finally, using

$$\frac{dv}{ds} = \frac{dz}{dt}\frac{1}{1-\alpha}s^{\frac{\alpha}{1-\alpha}}$$

and returning to the original variable t, we obtain the upper bound

$$\frac{1}{(1-\alpha)^2} \int_0^1 \left| \frac{dz}{dt} \right|^2 s^{\frac{2\alpha}{1-\alpha}}\,ds = \frac{1}{1-\alpha} \int_0^1 \left| \frac{dz}{dt} \right|^2 t^\alpha\,dt.$$

Case: $\alpha > 1$. Use change of variables $t^{\alpha-1} = s^{-1}$, proceed along the lines of the first case, using inequality $(4.4.38)_2$. See Exercise 4.4.1. $\qquad \square$

Lemma 4.18. *Let $\alpha \neq 0$, and let T be the master triangle. There exists a constant $C > 0$ such that, for all u for which*

$$\int_T |x|^\alpha |\nabla u|^2 < \infty,$$

there exists a constant a such that

$$\int_T |x|^{\alpha-2} |u - a|^2 \leq C \int_T |x|^\alpha |\nabla u|^2. \tag{4.4.40}$$

For $\alpha < 0$ and continuous functions u, $a = u(0)$.

Proof. *Step* 1: We first prove the result for the quadrant of the unit circle:

$$S := \{(r, \theta) : r < 1, \, 0 < \theta < \pi/2\}.$$

Consider the average of u in θ,

$$\bar{u}(r) = \frac{2}{\pi} \int_0^{\pi/2} u(r, \theta) \, d\theta.$$

Let $0 < r_1 < r_2 < 1$. We have

$$
\begin{aligned}
|\bar{u}(r_2) - \bar{u}(r_1)| &= \frac{2}{\pi} \left| \int_0^{\pi/2} (u(r_2, \theta) - u(r_1, \theta)) \, d\theta \right| \\
&= \frac{2}{\pi} \left| \int_0^{\pi/2} \int_{r_1}^{r_2} r^{-\frac{\alpha+1}{2}} r^{\frac{\alpha+1}{2}} \frac{\partial u}{\partial r} \, dr d\theta \right| \\
&\leq \frac{2}{\pi} \left(\frac{\pi}{2} \int_{r_1}^{r_2} r^{-(\alpha+1)} \, dr \right)^{1/2} \left(\int_0^{\pi/2} \int_{r_1}^{r_2} r^\alpha \left| \frac{\partial u}{\partial r} \right|^2 r dr \right)^{1/2} d\theta \\
&\leq \left(\frac{2}{\pi} \int_{r_1}^{r_2} r^{-(\alpha+1)} \, dr \right)^{1/2} \left(\int_S r^\alpha |\nabla u|^2 \, dS \right)^{1/2}.
\end{aligned}
$$

For $\alpha < 0$, integral $\int_0^1 r^{-(\alpha+1)} \, dr$ is finite, which implies that function $\bar{u}(r)$ is uniformly continuous and, therefore, admits a continuous extension to $r = 0$. For $\alpha > 0$, function $\bar{u}(r)$ is uniformly continuous in $[\epsilon, 1]$ for any $\epsilon > 0$.

We now have

$$
\begin{aligned}
\int_0^1 r^{\alpha+1} \left| \frac{d\bar{u}}{dr} \right|^2 dr &= \frac{4}{\pi^2} \int_0^1 r^{\alpha+1} \left| \int_0^{\pi/2} \frac{\partial u}{\partial r} \, d\theta \right|^2 dr \\
&\leq \frac{4}{\pi^2} \int_0^1 r^{\alpha+1} \left(\frac{\pi}{2} \right)^2 \int_0^{\pi/2} \left| \frac{\partial u}{\partial r} \right|^2 d\theta \, dr \\
&= \int_S r^\alpha \left| \frac{\partial u}{\partial r} \right|^2 dS \leq \int_S r^\alpha |\nabla u|^2 \, dS.
\end{aligned}
$$

Using Lemma 4.17, we obtain

$$\int_0^1 r^{\alpha-1} |\bar{u}(r) - a|^2 \, dr \leq C \int_0^1 r^{\alpha+1} \left| \frac{d\bar{u}}{dr} \right|^2 dr \leq C \int_S r^\alpha |\nabla u|^2 \, dS,$$

where $a = \bar{u}(0)$ for $\alpha < 0$, and $a = \bar{u}(1)$ for $\alpha > 0$. In addition, for $\alpha < 0$, if $u(r,\theta)$ is continuous on $\bar{S}$, then $a = u(0)$; compare Exercise 4.4.3. Integrating in θ, we get

$$\int_S r^{\alpha-2}|\bar{u} - a|^2 \, dS \leq C \int_S r^{\alpha}|\nabla u|^2 \, dS. \tag{4.4.41}$$

The Intermediate Value Theorem implies that there exists an angle ψ such that $\bar{u}(r) = u(r,\psi)$. Consequently,

$$u(r,\phi) - \bar{u}(r) = u(r,\phi) - u(r,\psi) = \int_\psi^\phi \frac{\partial u}{\partial \theta}(r,\theta) \, d\theta$$

$$\leq C \left[\int_0^{\pi/2} \left| \frac{\partial u}{\partial \theta}(r,\theta) \right|^2 d\theta \right]^{1/2}.$$

Integrating in ϕ,

$$\int_0^{\pi/2} |u(r,\phi) - \bar{u}(r)|^2 \, d\phi \leq C \int_0^{\pi/2} \left| \frac{\partial u}{\partial \theta}(r,\theta) \right|^2 d\theta.$$

Finally, multiplying both sides with $r^{\alpha-2}$ and integrating in r, we obtain

$$\int_0^1 r^{\alpha-2} \int_0^{\pi/2} |u(r,\phi) - \bar{u}(r)|^2 d\phi \, r dr \leq C \int_0^1 r^{\alpha-1} \int_0^{\pi/2} \left| \frac{\partial u}{\partial \theta} \right|^2 d\theta \, dr$$

$$= C \int_0^1 r^{\alpha} \int_0^{\pi/2} \left| \frac{1}{r}\frac{\partial u}{\partial \theta} \right|^2 d\theta \, r dr$$

$$\leq C \int_S r^{\alpha}|\nabla u|^2 \, dS.$$

Using the triangle inequality, estimate (4.4.41), and the estimate above, we get the required result.

Step 2: Consider the map from the master element into the section S,

$$\begin{cases} r' = \dfrac{1}{a(\theta)}r, \\ \theta' = \theta, \end{cases}$$

where $a(\theta)$ is defined in Fig. 4.5, and use the change of variables. □

Lemma 4.19. *Let $\epsilon > 0$, $0 < s < 1$. There exists a constant $C = C(\epsilon, s)$ such that, for every triangle T with vertices $v_0 = 0, v_1, v_2$, and a minimum angle $\geq \epsilon$, the following inequality holds:*

$$\int_T |x|^{2s-4}|u - p|^2 + |x|^{2s-2}|d_x^1(u - p)|^2 \leq C \int_T |x|^{2s}|d_x^2 u|^2 \tag{4.4.42}$$

for every function u such that

$$\int_T |u|^2 + |d_x^1 u|^2 + |x|^{2s}|d_x^2 u|^2 < \infty.$$

Here p stands for the vertex interpolant of u, and d_x^1, d_x^2 denote the first and second differentials of function u with norms

$$|d_x^1 u|^2 = |\nabla u|^2 = \sum_{|\alpha|=1} |D^\alpha u|^2, \qquad |d_x^2 u|^2 = \sum_{|\alpha|=2} |D^\alpha u|^2.$$

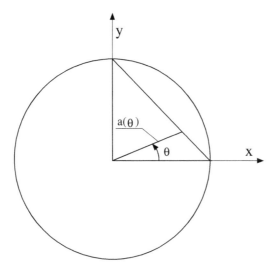

Figure 4.5: Mapping the master element into the quadrant of a circle.

Proof. Recall the earlier discussion on regularity of functions from the weighted Sobolev space to realize that functions u are continuous and, therefore, the vertex interpolant is well-defined. **Case:** T is the master triangle, $v_1 = (1, 0), v_2 = (0, 1)$. Set $\alpha = 2s$ in Lemma 4.18 to claim

$$\int_T |x|^{2s-2} \left| \frac{\partial u}{\partial x_i} - a_i \right|^2 \leq C \int_T |x|^{2s} \left| \nabla \left(\frac{\partial u}{\partial x_i} \right) \right|^2 \leq C \int_T |x|^{2s} |d_x^2 u|^2 \,.$$

Now replace u with $v = u - a_1 x_1 - a_2 x_2$, and use Lemma 4.18 with $\alpha = 2s - 2$ to obtain

$$\int_T |x|^{2s-4} |v - v(0)|^2 \leq C \int_T |x|^{2s-2} |\nabla v|^2 \,.$$

Combining the two estimates, we get estimate (4.4.42) but with vertex interpolant p replaced with polynomial $q = u(0) + a_1 x_1 + a_2 x_2$. In order to correct the polynomial, consider function $u_0 = u - q = v - v(0)$ and polynomial $p_0 = p - q$. Note that $p_0 = 0$ at $v_0 = 0$, and

$$p_0(v_1) = p_0((1, 0)) = u(v_1) - (u(0) + a_1) = u_0(v_1) \,.$$

Similarly, $p_0(v_2) = u_0(v_2)$. We have now

$$
\begin{aligned}
\int_T &|x|^{2s-4} |p_0|^2 + |x|^{2s-2} |d_x^1 p_0|^2 \\
&\leq C \left(|p_0(v_2)|^2 + |p_0(v_3)|^2 \right) && \text{(finite-dimensionality argument)} \\
&= C \left(|u_0(v_2)|^2 + |u_0(v_3)|^2 \right) \\
&\leq C \int_T |u_0|^2 + |d_x^1 u_0|^2 + |x|^{2s} |d_x^2 u_0|^2 && \text{(continuous embedding argument)} \\
&\leq C \int_T |x|^{2s} |d_x^2 u|^2 && \text{(estimate for function } u) \,.
\end{aligned}
$$

Use the triangle inequality and the estimates for $u_0 = u - q$ and $p_0 = p - q$ to arrive at the final estimate for $u_0 - p_0 = u - p$.

Case: Arbitrary triangle T. Use the linear map

$$x = B\xi$$

with a nonsingular matrix B, and a standard scaling argument; see Exercise 4.4.2. □

Theorem 4.20 (Babuška, Kellogg, and Pitkaränta [5]). *Let $\mathcal{T}$ be a triangulation of type* (h, γ, L). *The following interpolation error estimate holds:*

$$\|u - \Pi u\|_{H^1(\Omega)} \le Ch|u|_{H^{2,\gamma}(\Omega)}, \qquad u \in H^{2,\gamma}(\Omega) \cap H_D^1(\Omega), \tag{4.4.43}$$

where $C = C(\gamma, L)$, and Πu denotes the linear vertex interpolant of u.

Proof. We begin by recalling that space $H^{2,\gamma}(\Omega)$ is embedded in $C(\bar{\Omega})$ and, therefore, the vertex interpolant $v := \Pi_h u$ is well-defined.
Case: Element T without a singular vertex. The standard interpolation error estimate reads

$$\|u - v\|_{H^1(T)}^2 \le Cd_T^2 |u|_{H^2(T)}^2,$$

where constant C depends only upon the minimal angle, and element diameter d_T satisfies the condition

$$d_T \le Lh \inf_{x \in T} \phi_\gamma(x),$$

and, trivially,

$$L^2 h^2 \inf_{x \in T} \phi_\gamma^2(x) \int_T \sum_{|\alpha|=2} |D^\alpha u|^2 \le L^2 h^2 \int_T \phi_\gamma^2(x) \sum_{|\alpha|=2} |D^\alpha u|^2.$$

Consequently,

$$\|u - v\|_{H^1(T)}^2 \le Ch^2 |u|_{H^{2,\gamma}(T)}^2.$$

Case: Element T with a singular vertex x_i and weight $0 \le \gamma_i < 1$. Assume for simplicity that $x_i = 0$. For $x \in T$ and $s < 1$,

$$|x| \le d_T \qquad \Rightarrow \qquad |x|^{2(s-1)} \ge d_T^{2(s-1)},$$

so, by the interpolation estimate (4.4.42),

$$d_T^{2(s-1)} \int_T |u - v|^2 + |\nabla(u - v)|^2 \le C \int_T |x|^{2s} |d_x^2 u|^2.$$

Setting $s = \gamma_i$, we obtain

$$\int_T |u - v|^2 + |\nabla(u - v)|^2 \le Cd_T^{2(1-\gamma_i)} \int_T |x|^{2\gamma_i} |d_x^2 u|^2.$$

But, by the mesh design,

$$d_T \le Lh \sup_{x \in T} \phi_\gamma(x) \le Ch \sup_{x \in T} |x - x_i|^{\gamma_i} \le Chd_T^{\gamma_i},$$

so

$$d_T^{1-\gamma_i} \le Ch,$$

which yields the desired estimate. Summing up the element interpolation error estimates over all elements T, we obtain the global estimate. □

The mesh parameter h can be estimated by the total number of vertices N (d.o.f.).

Lemma 4.21. *There exists a constant $C > 0$, dependent upon Ω, γ, L but independent of h, such that, for every triangulation $\mathcal{T}$ of type (h, γ, L),*

$$N \leq Ch^{-2}. \tag{4.4.44}$$

Proof. Clearly,

$$N \leq 3\,\#\,\text{elements}.$$

As the number of elements adjacent to singular vertices is finite, it is sufficient to estimate the number of elements that are not adjacent to any of the singular vertices. By the mesh design, we have

$$L^{-1}h\phi_\gamma(x) \leq d_T \quad \Rightarrow \quad L^{-2}d_T^{-2} \leq h^{-2}\phi_\gamma^{-2}(x),$$

so

$$L^{-2}d_T^{-2}\underbrace{\int_T 1}_{=|T|} \leq h^{-2}\int_T \phi_\gamma^{-2}.$$

Shape regularity implies that there exists a constant C such that

$$1 \leq Cd_T^{-2}|T| = Cd_T^{-2}\int_T 1 \leq Ch^{-2}\int_T \phi_\gamma^{-2}.$$

Consequently,

$$\#\,\text{elements} \leq Ch^{-2}\underbrace{\int_\Omega \phi_\gamma^{-2}}_{<\infty}. \qquad \square$$

Lemma 4.21 implies that mesh parameter h in estimate (4.4.43) can be replaced with N^{-2}. Note that estimate (4.4.44) holds trivially for quasi-uniform meshes. Use of meshes graded according to the weight function ϕ_γ restores thus the optimal rate of convergence in terms of the total number of d.o.f. N.

Exercises

4.4.1. Prove the second case of Lemma 4.17 using the hint in the text. (3 points)

4.4.2. Provide the scaling arguments in the end of the proof of Lemma 4.19 to finish the proof for an arbitrary triangle adjacent to the origin. (3 points)

4.4.3. Let $u(r, \theta)$ be a continuous function in $\bar{S}$ where S is the first quadrant of the unit circle. Let $\bar{u}$ denote the average of function u in θ, i.e.,

$$\bar{u}(r) := \frac{2}{\pi}\int_0^{\pi/2} u(r, \theta)\,d\theta, \quad r > 0.$$

Prove that $\bar{u}$ is continuous in $(0, 1]$ and

$$\lim_{r \to 0} \bar{u}(r) = u(0).$$

(3 points)

Chapter 5

The DPG Method

The last chapter of the notes is devoted to an introductory exposition of the discontinuous Petrov–Galerkin (DPG) method with optimal test functions. We begin with the concept of an ideal Petrov–Galerkin method with optimal test functions and the corresponding practical realization of it. Next, we discuss the "breaking test spaces and forms" paradigm, i.e., the concept of variational formulations with discontinuous (broken) test spaces. We first use the grad-div problems and then extend the theory to curl-curl Maxwell problems. Construction of necessary Fortin operators is presented next, and we finish with the discussion of the double adaptivity approach.

5.1 ▪ The Ideal Petrov–Galerkin Method

The adventure with the DPG method, co-invented with Jay Gopalakrishnan, started quite a few years ago [32, 33], and I still have not converged to a unique way of presenting (and understanding) it. DPG stands for the *discontinuous Petrov–Galerkin* method, and the name was "stolen" from Italian colleagues who used it for what we later renamed the *ultraweak variational formulation* [9, 17]. The full name should be the *discontinuous Petrov–Galerkin method with optimal test functions*. The word *discontinuous* refers here to the use of discontinuous or *broken* test spaces only.[18] The method combines the fundamental concept of *Petrov–Galerkin discretization with optimal test functions* and the use of broken test spaces technologies that makes it a practical, implementable method within a standard Galerkin FE code supporting the exact sequence. To make it worse, there are the *ideal DPG method* and the *practical DPG method*. The word *ideal* refers to an idealized scenario where the optimal test functions are computed exactly. Except for 1D model problems, such a computation is not possible, and we have to approximate them using the good old Bubnov–Galerkin method and *enriched test spaces*. In other words, in practice, we always compute with the practical DPG method only. If all of this sounds complicated, there is still a concept of *double adaptivity* introduced by Wolfgang Dahmen and his collaborators [20] which is directly related to the ideal DPG method, but it does not fit the framework of the practical DPG method at all.

We start with our standard abstract variational formulation with a nonsymmetric functional setting,

$$
\begin{cases} u \in U, \\ b(u,v) = l(v), \quad v \in V, \end{cases}
\qquad \Leftrightarrow \qquad
\begin{cases} u \in U, \\ Bu = l, \end{cases}
$$

where, as usual, $B : U \to V'$ is the operator generated by the bilinear (sesquilinear) form.

---

[18]Some of my colleagues prefer to call them *product test spaces*.

The sesquilinear form $b(u, v)$ satisfies the inf-sup condition or, equivalently, operator B is bounded below. The Closed Range Theorem tells us that the boundedness below is a must if we want our problem to be well-posed. A direct discretization of the variational problem with the Petrov–Galerkin (PG) method leads to a pair of discrete spaces, the trial space $U_h \subset U$ and the test space $V_h \subset V$. They must be of equal dimension, $\dim U_h = \dim V_h =: N$, in order to obtain a system of N linear equations with N unknowns. Babuška's Theorem asks for a discrete counterpart of the inf-sup condition with a discrete inf-sup constant γ_h. This constant must be uniformly bounded away from zero,

$$\gamma_h \geq \gamma_0 > 0\,,$$

if we want to see the actual FE error and the *best approximation error* converge to zero with the same rates.

The practical question is now, *How do we select the discrete spaces?* The choice of trial space U_h is dictated by *approximability*. Given whatever information we can collect about the regularity of the anticipated solution, we want to select our trial space elements so they can approximate the unknown exact solution as well as possible. Historically, the FE business started with quasi-uniform meshes. As the rate of convergence is limited by both polynomial order *and regularity of the solution* (expressed in terms of Sobolev spaces), it made little sense to use higher order elements for irregular solutions, even if the lack of regularity was caused by isolated singularities. Later on, we learned that, in the case of isolated irregularities, the h-adaptivity could restore the optimal rates of convergence (as dictated by the polynomial degree), so this restriction in choosing the trial space element was removed.[19] Seeking exponential rates of convergence led to hp-adaptivity in the trial space and so on.

The story is much more complicated with the choice of the discrete test spaces. In the case of coercive problems, the stability is not an issue and we can stick with the Galerkin method. But what about the noncoercive problems? It has become gradually understood that the test spaces have to be selected with stability in mind. One of the early attempts to address the issue was the concept of optimal test functions (and spaces) by Barret and Morton [6]. We will discuss it in a moment. Jay's and my idea was different. We proposed using test functions that *realize the supremum in the inf-sup condition*, i.e., for each discrete trial function u_h, we want to find a corresponding *optimal test function* $v_u \in V$ such that

$$\sup_{v \in V} \frac{|b(u, v)|}{\|v\|_V} = \frac{|b(u, v_u)|}{\|v_u\|_V}\,.$$

Function v_u is sometimes called a *supremizer*. First of all, we knew right away that, in the reflexive Banach setting, v_u exists and is unique. This follows from standard weak compactness and strict convexity arguments. Thus, we can talk about *the* supremizer. If, by any luck, *trial-to-test* operator $T : U \ni u \to v_u \in V$ is linear, we can employ for an *optimal test space* the image of the trial space through the trial-to-test operator,

$$V_h^{\text{opt}} := T(U_h)\,.$$

With such an optimal test space, the continuous inf-sup condition *automatically implies* the satisfaction of the discrete inf-sup condition. Indeed,

$$\sup_{v_h \in V_h^{\text{opt}}} \frac{|b(u_h, v_h)|}{\|v_h\|_V} \geq \frac{|b(u_h, Tu_h)|}{\|Tu_h\|_V} = \sup_{v \in V} \frac{|b(u_h, v)|}{\|v\|_V} \geq \gamma \|u_h\|_U\,.$$

In other words, $\gamma_h \geq \gamma$.

---

[19]This remark does not apply to problems with solutions that are irregular "everywhere" like dynamic contact-impact problems. It is not a coincidence that the entire crashworthiness industry is using linear elements only.

Proposition 5.1. *The trial-to-test operator is defined by*

$$Tu = R_V^{-1} Bu, \qquad u \in U,$$

where $R_V : V \to V'$ is the Riesz operator corresponding to test inner product. In particular, T is indeed linear.

Proof. Recall that the Riesz operator is an isometric isomorphism from V onto its dual V'; see [66, p. 513]. We have

$$\sup_{v \in V} \frac{|b(u,v)|}{\|v\|_V} = \|b(u,\cdot)\|_{V'} = \|Bu\|_{V'} = \|R_V^{-1} Bu\|_V$$
$$= \frac{(R_V^{-1}Bu, R_V^{-1}Bu)_V}{\|R_V^{-1}Bu\|_V} = \frac{\langle Bu, Tu \rangle}{\|Tu\|_V} = \frac{b(u,Tu)}{\|Tu\|_V} = \frac{|b(u,Tu)|}{\|Tu\|_V},$$

as claimed. □

One of the immediate consequences of using the optimal test functions is the symmetry and positive definiteness of the PG stiffness matrix:

$$B_{ij} := b(e_j, \underbrace{Te_i}_{=:g_i}).$$

Proposition 5.2. *Stiffness matrix B_{ij} is Hermitian and positive definite,*

$$B_{ij} = \overline{B_{ji}} > 0.$$

Proof. Decoding the definition of the optimal test function $g_i = Te_i$ corresponding to trial basis function e_i, we have

$$g_i = Te_i = R_V^{-1}Be_i \quad \Leftrightarrow \quad R_V g_i = Be_i \quad \Leftrightarrow \quad \begin{cases} g_i \in V, \\ (g_i, \delta v)_V = b(e_i, \delta v), \quad \delta v \in V. \end{cases}$$

Consequently,

$$\begin{aligned} b(e_j, Te_i) &= (Te_j, Te_i) & \text{(definition of } Te_j) \\ &= \overline{(Te_i, Te_j)} & \text{(inner product is Hermitian)} \\ &= \overline{b(e_i, Te_j)} & \text{(definition of } Te_i). \end{aligned}$$

The positive definiteness of the test inner product and injectivity of the trial-to-test operator T imply that the stiffness matrix is positive definite as well,

$$B_{ij} = b(e_j, Te_i) = (Te_j, Te_i) > 0. \qquad □$$

The properties of the stiffness matrix indicate that the PG method with optimal test functions is, like least squares, a minimum residual method. This is indeed the case. In order to see that, we start by introducing a new, so-called energy norm in the trial space,

$$\|u\|_E := \|Tu\|_V = \|R_V^{-1}Bu\|_V = \|Bu\|_{V'}. \tag{5.1.1}$$

The energy norm is equivalent to the original norm in U with continuity constant M and inf-sup constant γ being the equivalence constants. Indeed,

$$\|u\|_E = \|Bu\|_{V'} \le M\|u\|_U \qquad \text{and} \qquad \gamma\|u\|_U \le \|Bu\|_{V'} = \|u\|_E .$$

If we replace the original norm in U with the energy norm, the corresponding new continuity and inf-sup constants are equal to one; compare Exercise 5.1.1. Note that the definition of optimal test functions has nothing to do with the norm in U, and therefore changing the norm in U does not affect the optimal test functions. Let u_h be the approximate solution obtained with the optimal test functions. We have

$$\|u - u_h\|_E \le \frac{1}{\gamma_h} \inf_{w_h \in U_h} \|u - w_h\|_E \qquad \text{(Babuška Theorem with } M = 1\text{)}$$
$$\le \inf_{w_h \in U_h} \|u - w_h\|_E \qquad (\gamma_h \ge \gamma = 1) .$$

As $u_h \in U_h$ itself, we must have the equality above, i.e.,

$$\|u - u_h\|_E = \inf_{w_h \in U_h} \|u - w_h\|_E .$$

In other words, the method delivers *the orthogonal projection in the energy norm.* Finally,

$$\|u - u_h\|_E = \|Bu - Bu_h\|_{V'} = \|l - Bu_h\|_{V'} ,$$

i.e., the energy norm of the error $e_h := u - u_h$, equals the residual measured in the dual norm.

The idea of optimal testing thus has led to a minimum residual method. Can we start with the minimum residual method and recover the optimal testing as well? The answer is "yes." Consider the minimum residual method,

$$J(u_h) = \min_{w_h \in U_h} J(w_h),$$
$$J(w_h) := \frac{1}{2}\|l - Bw_h\|_{V'}^2 = \frac{1}{2}\|R_V^{-1}(l - Bw_h)\|_V^2 = \frac{1}{2}\|R_V^{-1}(Bw_h - l)\|_V^2 , \qquad (5.1.2)$$

and compute the Gâteaux derivative of functional $J(w_h)$ at u_h,

$$\langle \partial J(u_h), \delta u_h \rangle = (R_V^{-1}(Bu_h - l), \underbrace{R_V^{-1}B\,\delta u_h}_{=T})_V ,$$

to obtain

$$(R_V^{-1}(Bu_h - l), \underbrace{R_V^{-1}B\,\delta u_h}_{=T})_V = 0, \quad \delta u_h \in U_h \quad \Leftrightarrow \quad b(u_h, T\delta u_h) = l(T\delta u_h), \quad \delta u_h \in U_h .$$

Note that we got rid of the inverse Riesz operators R_V^{-1} on the right by introducing the trial-to-test operator T and the one on the left by switching to the duality pairing. We can also do the opposite. Introducing the Riesz representation of the residual, $\psi := R_V^{-1}(l - Bu_h)$, we translate the optimality condition into the variational statement,

$$(\psi, R_V^{-1}B\delta u_h) = 0, \quad \delta u_h \in U_h \quad \Leftrightarrow \quad \overline{b(\delta u_h, \psi)} = 0, \quad \delta u_h \in U_h .$$

Decoding the definition of ψ,

$$\psi = R_V^{-1}(l - Bu_h) \quad \Leftrightarrow \quad R_V\psi + Bu_h = l \quad \Leftrightarrow \quad (\psi, v) + b(u_h, v) = l(v), \quad v \in V ,$$

we arrive at a special mixed problem,

$$\begin{cases} \psi \in V, \, u_h \in U_h \,, \\ (\psi, v) + b(u_h, v) = l(v), \quad v \in V \,, \\ \overline{b(\delta u_h, \psi)} = 0, \qquad \delta u_h \in U_h \,. \end{cases} \tag{5.1.3}$$

The mixed problem involves the approximate trial space U_h and the continuous, infinite-dimensional test space V. Note that the two Brezzi conditions are trivially satisfied. The (half-discrete) LBB condition is implied by the continuous inf-sup condition, and the inf-sup in kernel condition follows from the coercivity of the test norm. We have arrived at our final result in this section.

Theorem 5.3 (Three Hats of the Ideal PG Method). *The ideal Petrov–Galerkin method with optimal functions,*

$$\begin{cases} u_h \in U_h \,, \\ b(u_h, v_h) = l(v_h), \quad v_h \in V_h^{\mathrm{opt}} := TU_h \,, \end{cases} \tag{5.1.4}$$

the minimum residual method (5.1.2), and the mixed problem (5.1.3) are equivalent.

Finally, note that the method comes with *the built-in*[20] *residual a posteriori error estimate* $\|\psi\|_V$.

Exercises

5.1.1. Duality pairing. A bilinear (sesquilinear) form $b(u, v)$ defined on two Banach spaces U, V is called a *(generalized) duality pairing* if

$$\|u\|_U = \sup_{v \neq 0} \frac{|b(u, v)|}{\|v\|_V} \quad \text{and} \quad \|v\|_V = \sup_{u \neq 0} \frac{|b(u, v)|}{\|u\|_U} \,.$$

This implies that b must be *definite*, i.e.,

$$b(u, v) = 0 \quad \forall v \in V \quad \Rightarrow \quad u = 0 \,,$$
$$b(u, v) = 0 \quad \forall u \in U \quad \Rightarrow \quad v = 0 \,.$$

(i) Show that the standard duality pairing between a Banach space and its dual (with the induced norm) satisfies the axioms.

(ii) Let $b(u, v)$ be a continuous, definite form satisfying the inf-sup condition,

$$\gamma \|u\|_U \leq \sup_{v \neq 0} \frac{|b(u, v)|}{\|v\|_V} \,.$$

Replace the original norm in U with the *energy norm*:

$$\|u\|_E := \sup_{v \neq 0} \frac{|b(u, v)|}{\|v\|_V} \,.$$

Prove that the energy norm is indeed a norm on U and that with the energy norm replacing the original norm on U, form $b(u, v)$ becomes a duality pairing.

---

[20]In a standard adaptive FE method, we first solve the problem, and only *a posteriori* estimate the error. In the discussed PG method, we solve for the solution and the residual simultaneously.

(iii) Repeat the same argument with respect to v.

One arrives at nontrivial examples of duality pairings over the boundary of a domain when studying integration by parts and L^2-adjoints. (5 points)

5.1.2. Example of optimal test functions. Consider the classical variational formulation for a model 1D convection-dominated diffusion problem:

$$\begin{cases} u \in H_0^1(0,1), \\ \epsilon(u',v') + (u',v) = (f,v), \quad v \in H_0^1(0,1), \end{cases}$$

where $\epsilon > 0$ and $f \in L^2(0,1)$. Discretize the trial space with polynomials of order p,

$$U_p = \{u \in \mathcal{P}^p(0,1) : u(0) = u(1) = 0\} = \{u = x(1-x)w : w \in \mathcal{P}^{p-2}(0,1)\}.$$

Equip the test space with the H_0^1-norm,

$$\|v\|_V^2 := \|v'\|^2,$$

and determine analytically optimal test functions for the trial functions corresponding to polynomials $w = 1$ and $w = x$. (5 points)

5.2 ▪ The Practical Petrov–Galerkin Method

In practice, except for 1D model problems, we cannot determine the optimal test functions analytically and we have to somehow approximate them. Due to the symmetry and positive definiteness of the test inner product, approximation with the standard Bubnov–Galerkin method seems to be very natural. We introduce an *enriched test subspace* $V^r \subset V$, $\dim V^r \gg \dim U_h$, and compute the *approximate optimal test functions* using the standard Galerkin discretization,

$$\begin{cases} T^r u \in V^r, \\ (T^r u, \delta v)_V = b(u, \delta v), \quad \delta v \in V^r. \end{cases} \tag{5.2.5}$$

The *practical PG* method with optimal test functions is obtained by replacing the optimal test functions with approximate optimal test functions,

$$\begin{cases} \tilde{u}_h \in U_h, \\ b(\tilde{u}_h, T^r \delta u_h) = l(T^r \delta u_h), \quad \delta u_h \in U_h. \end{cases} \tag{5.2.6}$$

The operator $U_h \ni \delta u_h \to T^r(\delta u_h) \in V^r$ is termed *the approximate trial-to-test operator*. If we introduce the finite-dimensional Riesz operator corresponding to the enriched space,

$$R_{V^r} : V^r \to (V^r)' \quad \text{such that} \quad \langle R_{V^r} v, \delta v \rangle := (v, \delta v)_V, \quad \delta v \in V^r,$$

and the inclusion $\iota : V^r \hookrightarrow V$, the approximate trial-to-test operator can be represented as

$$T^r = R_{V^r}^{-1} \iota^T B.$$

It turns out that, as for the ideal PG method, the practical PG method is also equivalent to a minimum residual method and a mixed method. The residual is measured in the discrete dual norm induced by the enriched test space,

$$J(u_h) = \min_{w_h \in U_h} J(w_h), \qquad J(w_h) := \frac{1}{2}\|l - Bw_h\|_{(V^r)'}^2 = \frac{1}{2}\|R_{V^r}^{-1}\iota^T(l - Bw_h)\|_V^2. \tag{5.2.7}$$

Similarly, we have an an equivalent mixed method formulation:

$$\begin{cases} \psi^r \in V^r,\, \tilde{u}_h \in U_h, \\ (\psi^r, v^r) + b(\tilde{u}_h, v^r) = l(v^r), \quad v^r \in V^r, \\ \overline{b(\delta u_h, \psi^r)} = 0, \qquad \delta u_h \in U_h. \end{cases} \qquad (5.2.8)$$

We leave proving the following theorem to the reader (Exercise 5.2.1).

Theorem 5.4 (Three Hats of the Practical PG Method). *The Petrov–Galerkin method with approximate optimal functions,*

$$\begin{cases} u_h \in U_h, \\ b(u_h, v_h) = l(v_h), \quad v_h \in V_h^{r,\mathrm{opt}} := T^r U_h, \end{cases} \qquad (5.2.9)$$

the minimum residual method (5.2.7), and the mixed problem (5.2.8) are equivalent.

5.2.1 ▪ A Mixed Method Perspective

Once we have replaced the exact optimal test functions with the approximate test functions, we cannot claim anymore that the discrete inf-sup constant bounds the exact one. The supremum in the inf-sup condition is taken over a smaller, finite-dimensional enriched space and, in general, will be smaller than the supremum over the whole, infinite-dimensional test space. We must lose some stability and the question is, how much?

This is where the mixed method perspective turns out to be useful. We begin by reformulating the original problem, $Bu = l$, as the mixed problem:

$$\begin{cases} \psi \in V,\, u \in U, \\ R_V \psi + Bu = l, \\ B^* \psi = 0, \end{cases} \quad \Leftrightarrow \quad \begin{cases} \psi \in V,\, u \in U, \\ (\psi, v)_V + b(u, v) = l(v), \quad v \in V, \\ \overline{b(\delta u, \psi)} = 0, \qquad \delta u \in U. \end{cases} \qquad (5.2.10)$$

If the original problem is well-posed, i.e., form b satisfies the inf-sup condition, and form l satisfies the compatibility condition (possibly trivial), then, similarly to the discrete level, the original and mixed problems are equivalent to each other, i.e., u is a solution to $Bu = l$ iff the pair $(\psi = 0, u)$ is the solution to (5.2.10); see Exercise 5.2.2. Note also that the mixed problem may be well-posed even if form l does not satisfy the compatibility condition.

Once we have established the equivalence, the practical DPG mixed problem (5.2.8) can be viewed as a discretization of (5.2.10) and we can invoke Brezzi's theory to investigate its discrete stability and convergence. As for the classical Stokes problem, the inf-sup in kernel condition is satisfied trivially since the test inner product is coercive. The LBB inf-sup condition, at the first glance, seems to be simply the original discrete Babuška inf-sup condition,

$$\sup_{v^r \in V^r} \frac{|b(u_h, v^r)|}{\|v^r\|_V} \geq \gamma_h \|u_h\|_U.$$

It looks like we have come back to the starting point and gained nothing. This is not the case. Contrary to Babuška's Theorem where the discrete trial and test spaces must be of the same dimension, Brezzi's theory allows the (enriched) discrete test space V^r to have a bigger dimension than the trial space. Intuitively speaking, we may increase the dimension of the enriched test space until the (discrete) inf-sup condition is satisfied.

At the end of the day, we need to construct a Fortin operator, and we will show such a construction for the actual DPG method which employs *discontinuous or broken* test functions. As we will see, the broken test spaces make such constructions much easier than the globally conforming spaces in the classical setting. Once we construct the Fortin operator with a continuity constant C_F, we can claim the standard convergence result for the mixed problem:

$$\left(\|\psi - \psi^r\|_V^2 + \|u - u_h\|_U^2 \right)^{1/2} \leq C(M, \gamma, C_F) \left(\inf_{\phi^r \in V^r} \|\psi - \phi^r\|_V^2 + \inf_{w_h \in U_h} \|u - w_h\|_U^2 \right)^{1/2},$$

where the dependence of the ultimate stability constant upon M, γ, and C_F was discussed in Section 4.3. The critical fact about this special mixed problem is that the "exact" residual is zero, $\psi = 0$. Consequently, the corresponding best approximation error is zero as well, and the estimate above reduces to

$$\left(\|\psi^r\|_V^2 + \|u - u_h\|_U^2 \right)^{1/2} \leq C(M, \gamma, C_F) \inf_{w_h \in U_h} \|u - w_h\|_U.$$

The mixed method perspective has turned out to be critical in goal-oriented a posteriori error estimation and, in particular, has led to the DPG* method [34]. One should not forget though that the ultimate discrete mixed problem is first of all an approximation of the ideal mixed problem. This perspective has led to the idea of *double adaptivity*; see Section 5.7 and [31].

Exercises

5.2.1. Prove Theorem 5.4. (3 points)

5.2.2. Explain in what sense problem $Bu = l$ and mixed problem (5.2.10) are equivalent to each other. (2 points)

5.3 • The Discontinuous Petrov–Galerkin Method

We shall first discuss variational formulations with discontinuous test functions (broken test spaces) and then follow with the introduction of the ideal and practical DPG methods. All results presented in this section are reproduced from [16].

5.3.1 • Nonsymmetric Functional Settings

One of the immediate consequences of the concept of optimal test functions is a diminished importance of the symmetric functional setting. Let $\Omega \subset \mathbb{R}^n$ be a bounded Lipschitz domain with boundary Γ split into two disjoint parts Γ_u and Γ_σ. Consider a model Poisson problem,

$$\begin{cases} -\Delta u = f & \text{in } \Omega, \\ u = u_0 & \text{on } \Gamma_u, \\ \dfrac{\partial u}{\partial n} = \nabla u \cdot n = \sigma_0 & \text{on } \Gamma_\sigma. \end{cases} \quad (5.3.11)$$

Multiplying the PDE with a test function v, integrating over Ω, and integrating by parts, we obtain

$$\int_\Omega \nabla u \cdot \nabla v - \int_{\Gamma_u} (\nabla u \cdot n) v - \int_{\Gamma_\sigma} (\nabla u \cdot n) v = \int_\Omega f v.$$

We build the natural BC into the formulation by replacing flux $\nabla u \cdot n$ with boundary data σ_0 and moving it to the right-hand side. Concerning the boundary integral over Γ_u, we eliminate it by *not testing on* Γ_u, i.e., we assume that $v = 0$ on Γ_u. The usual argument is to observe the relation with the underlying minimization problem where the homogeneous BC on test function v is necessary. The combination $u + \epsilon v$ has to satisfy the essential BC which leads to the homogeneous BC on v on Γ_u. Alternatively, we can argue that we have to make this assumption if we want a symmetric functional setting. In the case of $u_0 = 0$, the trial and test spaces are then identical:

$$U = V := \{v \in H^1(\Omega) : v = 0 \text{ on } \Gamma_u\}.$$

But what if we do not care about the symmetry? Do we still have to make this assumption?

Digression: Postprocessing the boundary flux. And what if the boundary flux $\sigma_n := \nabla u \cdot n$ is our primary object of interest? How do we compute it once a Galerkin approximation u_h to u has been obtained? A tempting option of differentiating directly numerical solution u_h is mathematically wrong. We control the convergence of u_h to u in the H^1-norm, which implies that the convergence of ∇u_h to ∇u is controlled only in the L^2-norm. This *does not* imply convergence of $\sigma_{n,h} := \nabla u_h \cdot n$ to $\sigma_n = \nabla u \cdot n$ on the boundary in any norm at all. In fact, for an arbitrary $u \in H^1(\Omega)$, the boundary flux is ill-defined, and it is mathematically illegal. Fortunately, we do have some additional a priori information about the solution u. With the assumption $-\Delta u = f \in L^2(\Omega)$, the boundary flux is well-defined and lives in the dual of $H^{1/2}(\Gamma_u)$. This follows from the integration by parts formula,

$$\int_{\Gamma_u} \sigma_n v = \int_\Omega \nabla u \cdot \nabla V + \int_\Omega \Delta u\, V - \int_{\Gamma_\sigma} \nabla u \cdot n\, V = \int_\Omega \nabla u \cdot \nabla V - \int_\Omega f\, V - \int_{\Gamma_\sigma} \sigma_0\, V\,,$$

$$(5.3.12)$$

where $v \in H^{1/2}(\Gamma_u)$, and $V \in H^1(\Omega)$ is an arbitrary finite energy lift of v. The right-hand side vanishes for any $v \in H^1(\Omega)$, $v = 0$ on Γ_u and, therefore, is independent of a particular extension V. Consequently, it defines a linear and continuous functional on $H^{1/2}(\Gamma_u)$; see Exercise 5.3.1. This suggests replacing u with its FE approximation u_h and computing the corresponding *approximate flux* through a *mathematical postprocessing* formula:

$$\int_{\Gamma_u} \sigma_{n,h} v_h = \int_\Omega \nabla u_h \cdot \nabla V_h + \int_\Omega f\, V_h - \int_{\Gamma_\sigma} \sigma_0\, V_h\,,$$

$$(5.3.13)$$

where v_h is an arbitrary FE function on boundary Γ_u, and V_h is an arbitrary FE lift of v_h to the whole domain. Upon approximating σ_n within an appropriate discrete trial space,[21] we obtain an additional system of discrete equations to be solved for $\sigma_{n,h}$. The orthogonality property

$$\int_\Gamma (\sigma_n - \sigma_{n,h}) v_h = \int_\Omega \nabla(u - u_h) \cdot \nabla V_h \qquad \forall v_h$$

$$(5.3.14)$$

leads to the estimate

$$\|\sigma_n - \sigma_{n,h}\|_{H^{-1/2}(\Gamma)} \leq C \|u - u_h\|_{H^1(\Omega)}\,.$$

Convergence of solution u_h to u in the H^1 energy norm implies convergence of the postprocessed flux $\sigma_{n,h}$ to the exact flux σ_n in a weak, energy implied, norm $\tilde{H}^{-1/2}(\Gamma_u)$; see Exercise 5.3.2.

Remark 5.5. The whole discussion above could be rephrased using the terminology of normal traces for the $H(\text{div}, \Omega)$ energy space. Indeed, for $\Delta u \in L^2(\Omega)$, the gradient $\sigma = \nabla u$ lives in the $H(\text{div}, \Omega)$ space and it has a well-defined normal trace. Note that, consistent with the Normal Trace Theorem [27, Section 4.1], you cannot separate normal n from ∇u on the boundary. The only meaningful object is the normal component of the flux.

---

[21] And, in particular, ensuring the discrete inf-sup condition.

The main point we want to make here is that we *do not have to assume that test functions vanish on* Γ_u. It follows from the presented energy considerations that the flux σ_n can be identified as a separate, new unknown. Instead of solving for u first, and only then postprocessing for σ_n, we can formulate a meaningful variational formulation where we solve simultaneously for both u and $\sigma_n =: \hat{t}$,

$$
\begin{cases}
u \in H^1(\Omega), \hat{t} \in H^{-1/2}(\Gamma), \\
(\nabla u, \nabla v) - \langle \hat{t}, v \rangle_\Gamma = (f, v), & v \in H^1(\Omega), \\
\quad\quad u = u_0 & \text{on } \Gamma_u, \\
\quad\quad \hat{t} = \sigma_0 & \text{on } \Gamma_\sigma,
\end{cases}
\tag{5.3.15}
$$

where $\langle \cdot, \cdot \rangle_\Gamma$ denotes the $H^{-1/2}(\Gamma) \times H^{1/2}(\Gamma)$ duality pairing on boundary Γ. From now on, we shall consistently denote all unknowns on the boundary with "hats."

Remark 5.6. The prescribed flux σ_0 lives in $H^{-1/2}(\Gamma_\sigma)$. This is consistent with the definition of $H^{-1/2}(\Gamma_\sigma)$ as the space of restrictions of distributions from $H^{-1/2}(\Gamma)$ to Γ_σ. Let $\tilde{\sigma}_0 \in H^{-1/2}(\Gamma)$ denote a finite energy lift of σ_0 to the whole boundary. The difference

$$
\hat{t}_u := \hat{t} - \tilde{\sigma}_0
$$

lives in space $\tilde{H}^{-1/2}(\Gamma_u)$, and we can reformulate variational problem (5.3.15) in the following form:

$$
\begin{cases}
u \in H^1(\Omega), \hat{t}_u \in \tilde{H}^{-1/2}(\Gamma_u), \\
(\nabla u, \nabla v) - \langle \hat{t}_u, v \rangle_{\Gamma_u} = (f, v) + \langle \tilde{\sigma}_0, v \rangle_\Gamma, & v \in H^1(\Omega), \\
\quad\quad u = u_0 & \text{on } \Gamma_u.
\end{cases}
\tag{5.3.16}
$$

Note that the boundary pairing on the left is now defined on the subset Γ_u of the boundary only. This is mathematically correct since the "tilde" space $\tilde{H}^{-1/2}(\Gamma_u)$ is indeed the dual of $H^{1/2}(\Gamma_u)$. You might say that we have built the flux BC into the formulation. We prefer to use the first formulation (5.3.15) for a number of reasons. First, we avoid introducing the technical definition of the tilde spaces. Second, the formulation is consistent with the standard FE implementation of essential BC where we first project boundary data into the FE space, and then use the FE shape functions to lift the (projected) Dirichlet data. This is exactly what we do when implementing a finite element discretization of problem (5.3.15). We first project both BC data u_0 and σ_0 to the appropriate FE spaces, lift them to the whole boundary with FE shape functions, form the modified load vector, and solve a problem with homogeneous BC. In some sense, one could say that we use the second formulation on a discrete level.

Remark 5.7. We would like to reiterate a technical point mentioned above. For $t \in H^{-1/2}(\Gamma)$ and $u \in H^{1/2}(\Gamma)$ the boundary integral is understood in the sense of the duality pairing,

$$
\int_\Gamma tu = \langle t, u \rangle_{H^{-1/2}(\Gamma) \times H^{1/2}(\Gamma)},
$$

and is mathematically meaningful. Restrictions of t and u to a part of the boundary, $\Gamma_0 \subset \Gamma$, live in the corresponding spaces of restrictions,

$$
t|_{\Gamma_0} \in H^{-1/2}(\Gamma_0), \quad u|_{\Gamma_0} \in H^{1/2}(\Gamma_0).
$$

However, the integral over the part of the boundary, $\int_{\Gamma_0} tu$, makes no longer sense mathematically since spaces $H^{-1/2}(\Gamma_0)$ and $H^{1/2}(\Gamma_0)$ are *not* dual to each other.

5.3.2 ▪ Broken Test Spaces

Let $\mathcal{T}_h$ be any partition of domain Ω. In practice, we will use a finite element mesh. The *broken* or *product* $H^1(\mathcal{T}_h)$ energy space is defined as follows:

$$H^1(\mathcal{T}_h) := \{v = \{v_K\} : v_K \in H^1(K), \quad K \in \mathcal{T}_h\}. \tag{5.3.17}$$

If we test the PDE in model problem (5.3.11) with a broken test function $v \in H^1(\mathcal{T}_h)$ over an element K and integrate by parts, we obtain

$$(\boldsymbol{\nabla} u, \boldsymbol{\nabla} v)_K - \langle \boldsymbol{\nabla} u \cdot n, v \rangle_{\partial K} = (f, v), \qquad v \in H^1(K).$$

Consistent with our previous discussion, we identify the normal flux σ_n as *a new independent variable* $\hat{t}$. Summing up over all elements, we obtain

$$\underbrace{\sum_K (\boldsymbol{\nabla} u, \boldsymbol{\nabla} v)_K}_{=:(\boldsymbol{\nabla} u, \boldsymbol{\nabla}_h v)} - \underbrace{\sum_K \langle \hat{t}, v \rangle_{\partial K}}_{=:\langle \hat{t}, v \rangle_{\Gamma_h}} = \underbrace{\sum_K (f, v)_K}_{=(f, v)}, \qquad v \in H^1(\mathcal{T}_h).$$

The notation $\boldsymbol{\nabla}_h$ indicates that the gradient of v is computed *elementwise*. This is consistent with the definition of the broken test space. The new unknown, flux $\hat{t}$, comes from a new energy space defined on the mesh skeleton Γ_h consisting of all element boundaries; this new space is defined as follows:

$$H^{-1/2}(\Gamma_h) := \left\{ \hat{t} \in \prod_K H^{-1/2}(\partial K) : \exists \sigma \in H(\mathrm{div}, \Omega) \text{ such that} \right.$$
$$\left. \gamma_n(\sigma|_K) = \hat{t} \text{ on } \partial K, \quad K \in \mathcal{T}_h \right\}, \tag{5.3.18}$$

where γ_n denotes the normal trace operator. The definition reflects the condition that the flux $\hat{t}$ should be *single-valued* on the mesh skeleton. Note the subtle details: restriction of $\sigma \in H(\mathrm{div}, \Omega)$ to element K lives in $H(\mathrm{div}, K)$, and the corresponding normal trace lives in $H^{-1/2}(\partial K)$. Thus, it makes sense to equate $\hat{t}_K \in H^{-1/2}(\partial K)$ with $\gamma_n \sigma|_K$ and to couple $\hat{t}$ with broken test functions v since the coupling is done *elementwise*.

We can now introduce our new variational formulation with the broken test space.

$$\begin{cases} u \in H^1(\Omega),\ \hat{t} \in H^{-1/2}(\Gamma_h), \\ (\boldsymbol{\nabla} u, \boldsymbol{\nabla}_h v) - \langle \hat{t}, v \rangle_{\Gamma_h} = (f, v), & v \in H^1(\mathcal{T}_h), \\ u = u_0 & \text{on } \Gamma_u, \\ \hat{t} = \sigma_0 & \text{on } \Gamma_\sigma. \end{cases} \tag{5.3.19}$$

More broken and skeleton spaces. As we turn to other applications involving exact sequence energy spaces, we develop analogous definitions for the $H(\mathrm{curl}, \Omega)$ and $H(\mathrm{div}, \Omega)$ spaces. We begin with the definition of the respective broken energy spaces,

$$H(\mathrm{curl}, \mathcal{T}_h) := \prod_{K \in \mathcal{T}_h} H(\mathrm{curl}, K),$$

$$H(\mathrm{div}, \mathcal{T}_h) := \prod_{K \in \mathcal{T}_h} H(\mathrm{div}, K).$$

Elements from $H(\mathrm{div}, \mathcal{T}_h)$ can be coupled with functions from a new skeleton energy space,

$$H^{1/2}(\Gamma_h) := \left\{ \hat{u} \in \prod_K H^{1/2}(\partial K) : \exists u \in H^1(\Omega) \text{ such that } \gamma(u|_K) = \hat{u}_K \text{ on } \partial K, \quad K \in \mathcal{T}_h \right\},$$

$$\tag{5.3.20}$$

where γ denotes the trace operator. As before, the coupling is done elementwise,

$$\langle \hat{u}, \sigma \rangle_{\Gamma_h} := \sum_K \langle \gamma_n \sigma_K, \hat{u}_K \rangle_{\partial K}, \quad \hat{u} \in H^{1/2}(\Gamma_h), \; \sigma \in H(\mathrm{div}, \mathcal{T}_h) \,.$$

Notice that we take the freedom of writing the skeleton function $\hat{u}$ in the duality pairing first, even though elementwise it is the normal trace $\gamma_n \sigma_K$ that acts on trace $\hat{u}_K$.

With the new spaces in place, we can expand our portfolio of variational formulations with broken test spaces. In particular, we can now use ultraweak (UW) formulations for problems involving grad and div operators. Continuing with our model problem, we can rewrite it as a system of first order PDEs,

$$\begin{cases} \sigma - \boldsymbol{\nabla} u = 0 & \text{in } \Omega \,, \\ - \mathrm{div}\, \sigma = f & \text{in } \Omega \,, \\ \quad u = u_0 & \text{on } \Gamma_u \,, \\ \quad \sigma_n = \sigma_0 & \text{on } \Gamma_\sigma \,. \end{cases}$$

It is convenient to introduce the formalism of first order systems.

$$\mathsf{u} := (\sigma, u) \,,$$

$$A\mathsf{u} := (\sigma - \boldsymbol{\nabla} u, - \mathrm{div}\, \sigma) \,,$$

$$D(A) := \{ (\sigma, u) \in H(\mathrm{div}, \Omega) \times H^1(\Omega) : \gamma_n \sigma = 0 \text{ on } \Gamma_\sigma, \quad \gamma u = 0 \text{ on } \Gamma_u \} \,,$$

$$\mathsf{v} := (\tau, v) \,,$$

$$A^*\mathsf{v} := (\tau + \boldsymbol{\nabla} v, \mathrm{div}\, \tau) \,, \qquad D(A^*) = D(A) \,,$$

$$H_A(\Omega) := \{ \mathsf{u} \in L^2(\Omega) : A\mathsf{u} \in L^2(\Omega) \} = H(\mathrm{div}, \Omega) \times H^1(\Omega) \,,$$

$$H_{A^*}(\Omega) := \{ \mathsf{v} \in L^2(\Omega) : A^*\mathsf{v} \in L^2(\Omega) \} = H(\mathrm{div}, \Omega) \times H^1(\Omega) \,.$$

As the nonhomogeneous BC are taken into account through finite energy lifts and modification of the right-hand side, we can first focus on the case of homogeneous BC. The standard UW formulation looks as follows:

$$\begin{cases} \mathsf{u} \in L^2(\Omega) \,, \\ (\mathsf{u}, A^*\mathsf{v}) = (\mathsf{f}, \mathsf{v}), \quad \mathsf{v} \in D(A^*) \,, \end{cases} \tag{5.3.21}$$

where $\mathsf{f} = (0, f)$. If we decide to test with functions from the whole energy space $H_{A^*}(\Omega)$, we have to introduce new unknowns: traces $\hat{\mathsf{u}} = (\hat{\sigma}_n, \hat{u}) \in H^{-1/2}(\Gamma) \times H^{1/2}(\Gamma) =: \hat{U}$,

$$\begin{cases} \mathsf{u} \in L^2(\Omega), \hat{\mathsf{u}} = (\hat{\sigma}_n, \hat{u}) \in \hat{U} \,, \\ (\mathsf{u}, A^*\mathsf{v}) - \langle \hat{\mathsf{u}}, \mathsf{v} \rangle_\Gamma = (\mathsf{f}, \mathsf{v}), \quad \mathsf{v} \in H_{A^*(\Omega)} \,, \\ \quad \hat{\sigma}_n = 0 \qquad \text{on } \Gamma_\sigma \,, \\ \quad \hat{u} = 0 \qquad \text{on } \Gamma_u \,. \end{cases} \tag{5.3.22}$$

5.3.3 ▪ Well-Posedness of Broken Variational Formulations

We turn to a more abstract notation that will accommodate all possible variational formulations with broken test spaces. As usual, we start with a "standard" abstract variational problem,

$$\begin{cases} u \in U \,, \\ b(u, v) = l(v), \quad v \in V \,, \end{cases}$$

with the bilinear form satisfying the inf-sup condition with constant γ. We assume that the original bilinear form can be extended to a broken test (super)space $V(\mathcal{T}_h) \supset V$, i.e., we have $b(u, v)$, $u \in U, v \in V(\mathcal{T})$. Note that we are overloading symbol $b(u, v)$. Similarly, we assume that the original linear form can be extended to the broken test space as well, $l(v), v \in V(\mathcal{T})$, and overload symbol l. We postulate next the existence of a skeleton energy space $\hat{U}$ and another bilinear form $\langle \hat{u}, v \rangle_{\Gamma_h}$, $\hat{u} \in \hat{U}$, $v \in V(\mathcal{T}_h)$, that satisfy the following property:

$$v \in V \quad \Leftrightarrow \quad \langle \hat{u}, v \rangle_{\Gamma_h} = 0 \quad \forall \hat{u} \in \hat{U}. \tag{5.3.23}$$

This condition indicates that the traces are Lagrange multipliers for enforcing conformity of test functions. Consider the *broken variational formulation*,

$$\begin{cases} u \in U, \hat{u} \in \hat{U}, \\ b(u, v) + \langle \hat{u}, v \rangle_{\Gamma_h} = l(v), \qquad v \in V(\mathcal{T}_h). \end{cases} \tag{5.3.24}$$

Is the broken formulation well-posed? More precisely, does the modified bilinear form

$$b_{\mathrm{mod}}((u, \hat{u}), v) := b(u, v) + \langle \hat{u}, v \rangle_{\Gamma_h}$$

satisfy the inf-sup condition? If the answer is yes, what is the corresponding inf-sup constant?

The answer follows from the original reasoning of Franco Brezzi for mixed problems. Consider a pair $(u, \hat{u})$ and (overload symbol l to) define

$$l(v) := b_{\mathrm{mod}}((u, \hat{u}), v).$$

In order to show the inf-sup condition, we need to demonstrate that we control u and $\hat{u}$ in terms of l. Control of u is an immediate consequence of the inf-sup condition for form $b(u, v)$ and assumption (5.3.23),

$$\|u\|_U \leq \gamma^{-1} \sup_{v \in V} \frac{|b(u, v)|}{\|v\|_V} = \gamma^{-1} \sup_{v \in V} \frac{|b_{\mathrm{mod}}((u, \hat{u}), v)|}{\|v\|_V} = \gamma^{-1} \sup_{v \in V} \frac{|l(v)|}{\|v\|_V}$$

$$\leq \gamma^{-1} \sup_{v \in V(\mathcal{T}_h)} \frac{|l(v)|}{\|v\|_{V(\mathcal{T}_h)}} = \gamma^{-1} \|l\|_{V(\mathcal{T}_h)'}.$$

Once we control u, we can move term $b(u, v)$ to the right-hand side,

$$\langle \hat{u}, v \rangle_{\Gamma_h} = l(v) - b(u, v),$$

to get the estimate

$$\sup_{v \in V(\mathcal{T}_h)} \frac{|\langle \hat{u}, v \rangle_{\Gamma_h}|}{\|v\|_{V(\mathcal{T}_h)}} \leq \|l\|_{V(\mathcal{T}_h)'} + M\|u\|_U \leq \left(1 + \frac{M}{\gamma}\right) \|l\|_{V(\mathcal{T}_h)'}. \tag{5.3.25}$$

Now, the question is whether the left-hand side is in fact a norm in which we can measure the Lagrange multiplier and, if the answer is yes, whether we can represent it in a more constructive way.

Before we answer this question in the abstract setting, we first consider the model problem. We have to unpack the abstract notation and go back to the concrete broken space setting. For the discussed model problem, we have

$$\hat{\mathsf{u}} = (\hat{u}, \hat{\sigma}_n) \in H^{1/2}(\Gamma_h) \times H^{-1/2}(\Gamma_h),$$

$$\mathsf{v} = (\tau, v) \in H(\mathrm{div}, \mathcal{T}_h) \times H^1(\mathcal{T}_h),$$

$$\langle \hat{\mathsf{u}}, \mathsf{v} \rangle_{\Gamma_h} = \langle \hat{u}, \tau_n \rangle_{\Gamma_h} + \langle \hat{\sigma}_n, v \rangle_{\Gamma_h}$$

and, trivially,

$$\left(\sup_{v}\frac{|\langle \hat{u}, v\rangle_{\Gamma_h}|}{\|v\|}\right)^2 = \left(\sup_{\tau\in H(\mathrm{div},\mathcal{T}_h)}\frac{|\langle \hat{u}, \tau_n\rangle_{\Gamma_h}|}{\|\tau\|_{H(\mathrm{div},\mathcal{T}_h)}}\right)^2 + \left(\sup_{v\in H^1(\mathcal{T}_h)}\frac{|\langle \hat{\sigma}_n, v\rangle_{\Gamma_h}|}{\|v\|_{H^1(\mathcal{T}_h)}}\right)^2.$$

It is a unique property of the broken test space that the supremum over the whole space (squared) is equal to the sum of the suprema over elements (squared),[22]

$$\left(\sup_{\tau\in H(\mathrm{div},\mathcal{T}_h)}\frac{|\langle \hat{u}, \tau_n\rangle_{\Gamma_h}|}{\|\tau\|_{H(\mathrm{div},\mathcal{T}_h)}}\right)^2 = \sum_K \left(\sup_{\tau\in H(\mathrm{div},K)}\frac{|\langle \hat{u}_K, \tau_n\rangle_{\partial K}|}{\|\tau\|_{H(\mathrm{div},K)}}\right)^2,$$

$$\left(\sup_{v\in H^1(\mathcal{T}_h)}\frac{|\langle \hat{\sigma}_n, v\rangle_{\Gamma_h}|}{\|v\|_{H^1(\mathcal{T}_h)}}\right)^2 = \sum_K \left(\sup_{v\in H^1(K)}\frac{|\langle \hat{\sigma}_{K,n}, v\rangle_{\partial K}|}{\|v\|_{H^1(K)}}\right)^2.$$

Thus, we can focus on the interpretation of the contribution from a single element K,

$$\sup_{\tau\in H(\mathrm{div},K)}\frac{|\langle \hat{u}_K, \tau_n\rangle_{\partial K}|}{\|\tau\|_{H(\mathrm{div},K)}} = \|\langle \hat{u}_K, \cdot\rangle_{\partial K}\|_{(H(\mathrm{div},K))'}.$$

Recalling the Riesz Theorem, it is sufficient to solve the variational problem,

$$\begin{cases} \tau \in H(\mathrm{div}, K), \\ (\tau, \delta\tau)_{H(\mathrm{div},K)} = \langle \hat{u}_K, \delta\tau_n\rangle_{\partial K}, & \delta\tau \in H(\mathrm{div}, K), \end{cases}$$

and compute the $H(\mathrm{div})$-norm of solution τ. This leads to the following Neumann boundary-value problem for τ:

$$\begin{cases} -\boldsymbol{\nabla}(\mathrm{div}\,\tau) + \tau = 0 & \text{in } K, \\ \mathrm{div}\,\tau = \hat{u} & \text{on } \partial K. \end{cases} \tag{5.3.26}$$

Lemma 5.8. *Let τ be the solution to Neumann problem (5.3.26). Then $u = \mathrm{div}\,\tau \in H^1(K)$ is the solution to the corresponding Dirichlet problem,*

$$\begin{cases} -\mathrm{div}(\boldsymbol{\nabla} u) + u = 0 & \text{in } K, \\ u = \hat{u} & \text{on } \partial K. \end{cases} \tag{5.3.27}$$

Moreover, $\|\tau\|_{H(\mathrm{div},K)} = \|u\|_{H^1(K)}$.

Proof. It is sufficient to apply the divergence operator to (5.3.26)$_1$. The equality of norms follows from the fact that $\tau = \boldsymbol{\nabla}(\mathrm{div}\,\tau) = \boldsymbol{\nabla} u$,

$$\|\mathrm{div}\,\tau\|^2 + \|\tau\|^2 = \|u\|^2 + \|\boldsymbol{\nabla}(\mathrm{div}\,\tau)\|^2 = \|u\|^2 + \|\boldsymbol{\nabla} u\|^2. \qquad \square$$

In conclusion,

$$\|\langle \hat{u}_K, \cdot\rangle_{\partial K}\|_{(H(\mathrm{div},K))'} = \|\hat{u}_K\|_{H^{1/2}(\partial K)},$$

where fractional space $H^{1/2}(\partial K)$ is equipped with the minimum energy extension norm:

$$\|\hat{u}\|_{H^{1/2}(\partial K)} = \inf_{\substack{u \in H^1(K) \\ u|_{\partial K} = \hat{u}}} \|u\|_{H^1(K)}.$$

---

[22]This implies, among other things, that the global residual (squared) equals the sum of element residuals (squared).

Similarly, application of the Riesz Theorem to the computation of the dual norm,

$$\sup_{v \in H^1(K)} \frac{|\langle \hat\sigma_{K,n}, v \rangle_{\partial K}|}{\|v\|_{H^1(K)}} = \|\langle \hat\sigma_{K,n}, \cdot \rangle_{\partial K}\|_{(H^1(K))'} \,,$$

leads to a variational problem for $u \in H^1(K)$,

$$(u, \delta u)_{H^1(K)} = \langle \hat\sigma_{K,n}, \delta u \rangle_{\partial K}, \qquad \delta u \in H^1(K) \,,$$

and, in turn, to the Neumann problem for Riesz representation $u \in H^1(K)$,

$$\begin{cases} -\operatorname{div}(\boldsymbol\nabla u) + u = 0 & \text{in } K \,, \\[4pt] \dfrac{\partial u}{\partial n} = \hat\sigma_{K,n} & \text{on } \partial K \,. \end{cases} \tag{5.3.28}$$

Lemma 5.9. *Let u be the solution to Neumann problem* (5.3.28). *Then* $\tau = \boldsymbol\nabla u \in H(\operatorname{div}, K)$ *is the solution to the corresponding Dirichlet problem,*

$$\begin{cases} -\boldsymbol\nabla(\operatorname{div}\tau) + \tau = 0 & \text{in } K \,, \\[4pt] \gamma_n(\tau) = \hat\sigma_{K,n} & \text{on } \partial K \,. \end{cases} \tag{5.3.29}$$

Moreover, $\|u\|_{H^1(K)} = \|\tau\|_{H(\operatorname{div}, K)}$.

Proof. It is sufficient to apply the gradient operator to (5.3.28)$_1$. The equality of norms follows from the fact that $u = \operatorname{div}(\boldsymbol\nabla u) = \operatorname{div}\tau$,

$$\|\boldsymbol\nabla u\|^2 + \|u\|^2 = \|\boldsymbol\nabla u\|^2 + \|\operatorname{div}(\boldsymbol\nabla u)\|^2 = \|\tau\|^2 + \|\operatorname{div}\tau\|^2 . \qquad \square$$

Similarly to the previous case, the dual norm of functional $\langle \hat\sigma_{K,n}, \cdot \rangle_{\partial K}$ turns out to be the minimum energy extension norm in $H^{-1/2}(\partial K)$.

In conclusion, for the considered model problem, the supremum on the left-hand side of (5.3.25) is indeed a norm, and it equals the minimum energy extension norm of traces $(\hat u, \hat\sigma_n)$. It is important to emphasize that the minimum energy extension norm for traces derives entirely from the employed test norm for the broken test space. Returning to the abstract setting, we assume that the norm for the broken test space $V(\mathcal T_h)$ is given in the form

$$\|v\|^2_{V(\mathcal T_h)} = \sum_{K \in \mathcal T_h} \|v_K\|^2_{V(K)} = \sum_K (\|Cv\|^2 + \|v\|^2) \,, \tag{5.3.30}$$

where C is a well-defined operator on group variable v. For the model problem, $C(\tau, v) = (\operatorname{div}\tau, \boldsymbol\nabla u)$. Computation of the supremum in (5.3.25) follows the same steps as for the model problem,

$$\left(\sup_{v \in V(\mathcal T_h)} \frac{|\langle \hat u, v \rangle|}{\|v\|_{V(\mathcal T_h)}} \right)^2 = \sum_K \left(\sup_{v \in V(K)} \frac{|\langle \hat u_K, v \rangle_{\partial K}|}{\|v\|_{V(K)}} \right)^2 .$$

Let $v \in V(K)$ now be the Riesz representation of the functional $\langle \hat u_K, v \rangle_{\partial K}$,

$$(v, \delta v)_{V(K)} = \langle \hat u_K, \delta v \rangle_{\partial K} \quad \forall \delta v \in V(K) \,.$$

The variational problem above translates into a Neumann problem for v,

$$\begin{cases} C^* C v + v = 0 & \text{in } K \,, \\[4pt] \gamma_{C^*}(Cv) = \hat u_K & \text{on } \partial K \,, \end{cases} \tag{5.3.31}$$

where γ_{C^*} is an appropriate trace operator. Similar to the reasoning in the two previous lemmas, this leads to a Dirichlet problem for $U = Cv$,

$$\begin{cases} CC^*U + U = 0 & \text{in } K \,, \\ \qquad \gamma_{C^*}U = \hat{u}_K & \text{on } \partial K \,. \end{cases} \qquad (5.3.32)$$

We can now better characterize the abstract trace space $\hat{U}$. We use the adjoint operator C^* to define a new energy space,

$$H_{C^*}(\Omega) := \{U \in L^2(\Omega) \,:\, C^*U \in L^2(\Omega)\} \,, \qquad (5.3.33)$$

along with the corresponding space of traces and trace operator,

$$\gamma_{C^*} \,:\, H_{C^*}(\Omega) \to \mathrm{tr}H_{C^*}(\Omega) \,. \qquad (5.3.34)$$

The abstract trace space $\hat{U}$ can be characterized in terms of element traces,

$$\hat{U} := \left\{ \hat{u} \in \prod_K \gamma_{C^*}H_{C^*}(K) \,:\, \exists U \in H_{C^*}(\Omega) \text{ such that } \gamma_{C^*}U|_K = \hat{u} \text{ on } \partial K, \quad K \in \mathcal{T}_h \right\} \,. \qquad (5.3.35)$$

According to the derivations above, the traces should be measured in the minimum energy extension norm. All of these definitions are purely formal,[23] but they clearly indicate that the functional setting for traces derives completely from the definition of the norm used for the broken test space and not from the trial norm on solution space U.

We conclude this section with our major result concerning the well-posedness of variational formulations with broken test spaces.

Theorem 5.10. *Let $V(\mathcal{T}_h)$ be a broken (product) test space with an inner product given by (5.3.30). Let $H_{C^*}(\Omega)$ be the corresponding energy space defined using (5.3.33), with the trace operator (5.3.34). Let $\hat{U}$ denote the mesh skeleton space, defined by (5.3.35), and equipped with the minimum energy extension norm. Assume that the broken test space contains a conforming subspace V and that property (5.3.23) holds.*

Let U be another Hilbert trial space, and let $b(u, v)$, $u \in U, v \in V(\mathcal{T}_h)$, be a bilinear (sesquilinear) form with continuity constant M. Assume that the restriction $b(u, v)$, $u \in U, v \in V$, satisfies the inf-sup condition with constant γ.

Then the modified bilinear form

$$b_{\mathrm{mod}}((u, \hat{u}), v) := b(u, v) + \langle \hat{u}, v \rangle_{\Gamma_h}$$

admits continuity constant $M_{\mathrm{mod}} \leq (M^2 + 1)^{1/2}$ and satisfies the inf-sup condition,

$$\sup_{v \in V(\mathcal{T}_h)} \frac{|b_{\mathrm{mod}}((u, \hat{u}), v)|}{\|v\|_{V(\mathcal{T}_h)}} \geq \gamma_{\mathrm{mod}}(\|u\|_U^2 + \|\hat{u}\|_{\hat{U}}^2)^{1/2} \,,$$

where γ_{mod} admits the lower bound:

$$\gamma_{\mathrm{mod}}^2 \geq \left(\frac{1}{\gamma^2} + \left(1 + \frac{M}{\gamma}\right)^2 \right)^{-1} \,.$$

---

[23]Pending a study of the energy space $H_{C^*}(\Omega)$ and its traces.

Remark 5.11. Test variable v is usually a group variable which makes C a vector-valued operator. If each component of Cv involves a single differential operator of grad, curl, or div, so does its adjoint C^*, and the abstract energy space $H_{C^*}(\Omega)$ reduces to products of standard energy spaces $H^1(\Omega), H(\mathrm{curl}, \Omega), H(\mathrm{div}, \Omega), L^2(\Omega)$ with the corresponding standard trace operators; see Exercise 5.3.3.

Implementation of the DPG method. We have finally arrived at the main point of using broken test spaces. Consider a general abstract broken variational formulation (5.3.24). The corresponding DPG method is based on discretizing the mixed problem:

$$\begin{cases} \psi \in V(\mathcal{T}_h), u \in U, \hat{u} \in \hat{U}, \\ (\psi, v)_V + b(u, v) + \langle \hat{u}, v \rangle_{\Gamma_h} = l(v), & v \in V(\mathcal{T}_h), \\ \qquad\qquad\quad b(\delta u, \psi) = 0, & \delta u \in U, \\ \qquad\qquad\quad \langle \delta \hat{u}, \psi \rangle_{\Gamma_h} = 0, & \delta \hat{u} \in \hat{U}. \end{cases} \tag{5.3.36}$$

Note that the residual $\psi = 0$. The *ideal DPG method* introduces discrete trial subspaces $U_h \subset U, \hat{U}_h \subset \hat{U}$ but leaves the exact test space untouched,

$$\begin{cases} \psi^h \in V(\mathcal{T}_h), u_h \in U_h, \hat{u}_h \in \hat{U}_h, \\ (\psi^h, v)_V + b(u_h, v) + \langle \hat{u}_h, v \rangle_{\Gamma_h} = l(v), & v \in V(\mathcal{T}_h), \\ \qquad\qquad\quad b(\delta u_h, \psi^h) = 0, & \delta u_h \in U_h, \\ \qquad\qquad\quad \langle \delta \hat{u}_h, \psi^h \rangle_{\Gamma_h} = 0, & \delta \hat{u}_h \in \hat{U}_h. \end{cases} \tag{5.3.37}$$

Finally, the *practical DPG method* discretizes the test space with an *enriched test space* $V_h(\mathcal{T}_h) \subset V(\mathcal{T}_h)$, to arrive at the final, fully discrete system:

$$\begin{cases} \psi_h \in V_h(\mathcal{T}_h), u_h \in U_h, \hat{u}_h \in \hat{U}_h, \\ (\psi_h, v_h)_V + b(u_h, v_h) + \langle \hat{u}_h, v_h \rangle_{\Gamma_h} = l(v_h), & v_h \in V_h(\mathcal{T}_h), \\ \qquad\qquad\quad b(\delta u_h, \psi_h) = 0, & \delta u_h \in U_h, \\ \qquad\qquad\quad \langle \delta \hat{u}_h, \psi_h \rangle_{\Gamma_h} = 0, & \delta \hat{u}_h \in \hat{U}_h. \end{cases} \tag{5.3.38}$$

System (5.3.38) translates into the system of linear equations

$$\begin{pmatrix} G & B & \hat{B} \\ B^T & 0 & 0 \\ \hat{B}^T & 0 & 0 \end{pmatrix} \begin{pmatrix} \psi \\ u \\ \hat{u} \end{pmatrix} = \begin{pmatrix} l \\ 0 \\ 0 \end{pmatrix},$$

where (overloaded) symbols $\psi, u, \hat{u}, l$ represent vectors of d.o.f. corresponding to residual ψ, solution u, Lagrange multipliers $\hat{u}$, and load l. For broken test space $V(\mathcal{T}_h)$ and the corresponding inner product,

$$(v, \delta v)_V = \sum_{K \in \mathcal{T}_h} (v_K, \delta v_K)_{V(K)},$$

Gram matrix G *is block-diagonal* which enables elementwise static condensation of residual ψ. This leads to the Schur complement for the remaining unknowns $u, \hat{u}$,

$$\begin{pmatrix} B^T G^{-1} B & B^T G^{-1} \hat{B} \\ \hat{B}^T G^{-1} B & \hat{B}^T G^{-1} \hat{B} \end{pmatrix} \begin{pmatrix} u \\ \hat{u} \end{pmatrix} = \begin{pmatrix} B^T G^{-1} l \\ \hat{B}^T G^{-1} l \end{pmatrix}.$$

In practice, we follow immediately with the static condensation of all *interior d.o.f.* of the unknown u. In particular, for the UW formulation, all d.o.f. belong to the element interior and can be condensed out elementwise. It is interesting to note that the number of interface d.o.f. is independent of the choice of variational formulation. The cost of the global solve (for the interface d.o.f.) is thus identical for all variational formulations which differ only in the cost of elementwise operations. Once unknowns $u, \hat{u}$ are determined, we follow with a second loop through elements to determine residual ψ.

A priori error estimate. Well-posedness of the broken variational formulation, ensured by Theorem 5.10, and the fact that the ideal DPG method reproduces the stability of the continuous problem, imply the a priori error bound for the ideal DPG method,

$$\left(\|u - u_h\|_U^2 + \|\hat{u} - \hat{u}_h\|_{\hat{U}}^2 \right)^{1/2} \leq C \left(\inf_{w_h \in U_h} \|u - w_h\|_U^2 + \inf_{\hat{w}_h \in \hat{U}_h} \|\hat{u} - \hat{w}_h\|_{\hat{U}}^2 \right)^{1/2} \quad (5.3.39)$$

with mesh-independent constant C. The same error bound will hold for the practical DPG method, provided we construct a Fortin operator with a mesh-independent continuity constant. The bound dictates the choice of discretization spaces for individual components of the unknown u and the trace $\hat{u}$. As for any mixed method, we choose the polynomial orders for the components of u and $\hat{u}$ in such a way that the corresponding best approximation errors (interpolation errors in practice) converge with the same rate. If the functional setting involves the exact sequence energy spaces, this implies that we should select the spaces forming the first exact sequence spaces discussed in Section 3.2. Note that this philosophy extends to the trace variables. For each element K,

$$\inf_{\hat{w}_h \in \hat{U}_h} \|\hat{u} - \hat{w}_h\|_{H^{1/2}(\partial K)} \leq \|U - \Pi_h^{\text{grad}} U\|_{H^1(K)} \leq C h_K^p \|U\|_{H^{p+1}(K)} ,$$

where U is any extension of the exact trace $\hat{u}$ (in practice the exact solution), and $\hat{U}_h$ is the trace of an appropriate H^1-conforming element of order p. Similarly,

$$\inf_{\hat{w}_h \in \hat{U}_h} \|\hat{v} - \hat{w}_h\|_{H^{-1/2}(\partial K)} \leq \|V - \Pi_h^{\text{div}} V\|_{H(\text{div},K)} \leq C h_K^p \|V\|_{H^{p+1}(\text{div},K)} ,$$

where, this time, $\hat{U}_h$ is the trace of the appropriate $H(\text{div})$-conforming element of order p, and V is an extension of trace $\hat{v}$.

By employing traces of H^1- and $H(\text{div})$-conforming elements to discretize the exact traces, we can implement the DPG method within any standard Galerkin FE code supporting the exact sequence. The whole discussion extends to the $H(\text{curl})$ energy space and its traces discussed in the next section.

Exercises

5.3.1. Prove that the right-hand side of (5.3.12) is a continuous functional of $V \in H^1(\Omega)$ and it is independent of extension V. Use the Trace Theorem to conclude that it defines a linear and continuous functional on the trace space $H^{1/2}(\Gamma_u)$. (3 points)

5.3.2. Flux postprocessing. Consider the model 2D Poisson problem (5.3.11) set up in the unit square domain Ω shown in Fig. 5.1. Consider a uniform mesh of bilinear elements and element size h. Let W_h denote the corresponding H^1-conforming FE mesh, and let u_h be

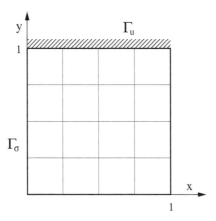

Figure 5.1: Model Poisson problem in a unit square domain.

the corresponding FE solution obtained using the standard Bubnov–Galerkin method,

$$
\begin{cases}
u_h \in W_{D,h}\,, \\
\displaystyle\int_\Omega \boldsymbol{\nabla} u_h \boldsymbol{\nabla} w_h = \int_\Omega f w_h + \int_{\Gamma_\sigma} \sigma_0 w_h\,, \qquad w_h \in W_{D,h}\,,
\end{cases}
\tag{5.3.40}
$$

where $W_{D,h}$ denotes the subspace of FE functions satisfying the Dirichlet BC,

$$
W_{D,h} := \{w_h \in W_h \,:\, w_h = 0 \text{ on } \Gamma_u\}\,.
$$

Let V_h denote the trace space of W_h on boundary Γ_u, and let U_h be the trace of the $H(\mathrm{div})$-conforming Raviart–Thomas space of the lowest order on boundary Γ_u, i.e., the space of piecewise constant functions; see Fig. 5.2. Consider the following postprocessing for flux $\sigma_{n,h}$ on Dirichlet boundary Γ_u:

$$
\begin{cases}
\sigma_{n,h} \in U_h\,, \\
\displaystyle\int_{\Gamma_u} \sigma_{n,h} v_h = \int_\Omega \boldsymbol{\nabla} u_h \boldsymbol{\nabla} w_h - \int_\Omega f w_h - \int_{\Gamma_\sigma} \sigma_0\, w_h\,, \quad v_h \in V_h\,,
\end{cases}
\tag{5.3.41}
$$

where $w_h \in W_h$ is an arbitrary FE extension of $v_h \in V_h$. Note that the test space is one dimension larger than the trial space and the system of equations must be solved in a minimum residual setting.

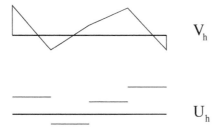

Figure 5.2: Discrete trial and test spaces of lowest order for flux reconstruction on Γ_u.

1. Explain why in (5.3.41) we can test with (piecewise) linear test functions but we cannot test with higher order (e.g., quadratic) test functions.

2. Recall Exercise 4.3.6 and explain why this "natural idea" for postprocessing is doomed to fail.

3. Recall Exercise 4.3.7 and correct the trial space for the postprocessed flux. Conclude by showing that there exists a mesh-independent constant C such that

$$\|\sigma_n - \sigma_{n,h}\|_{\tilde{H}^{-1/2}(\Gamma_u)} \leq C\|u - u_h\|_{H^1(\Omega)} \leq Ch^s\|u\|_{H^{1+s}(\Omega)}.$$

(5 points)

5.3.3. Breaking forms and test spaces in formulations for linear elasticity. Visit all formulations discussed in Section 1.4.2, and write out the corresponding formulations with broken test spaces eligible for the DPG method. (10 points)

5.4 ▪ Extension to Maxwell Problems

In this section, we discuss the use of broken $H(\operatorname{curl}, \mathcal{T}_h)$ test spaces in the context of Maxwell's equations. The general approach is identical to that discussed in Section 5.3 and leading to the formulation and proof of Theorem 5.10. The main technicality lies in the duality of the tangential trace with the rotated tangential trace.

Recall that the energy space $H(\operatorname{curl}, \Omega)$ comes with the *tangential trace* operator,

$$\gamma_t : H(\operatorname{curl}, \Omega) \ni E \to E_t \in H^{-1/2}(\operatorname{curl}_\Gamma, \Gamma).$$

The trace space $H^{-1/2}(\operatorname{curl}_\Gamma, \Gamma)$ is equipped with the minimum energy extension norm. The integration by parts formula,

$$\langle n \times E_t, F_t \rangle = (\boldsymbol{\nabla} \times E, F) - (E, \boldsymbol{\nabla} \times F),$$

motivates introducing the *rotated tangential trace* operator,

$$\gamma_t^\perp : H(\operatorname{curl}, \Omega) \ni E \to n \times E_t \in (H^{-1/2}(\operatorname{curl}_\Gamma, \Gamma))'.$$

The dual space $(H^{-1/2}(\operatorname{curl}_\Gamma, \Gamma))'$ is named $H^{-1/2}(\operatorname{div}_\Gamma, \Gamma)$ since, for a smooth manifold Γ, it indeed coincides with that space. For a C^1-manifold,

$$- \operatorname{div}_\Gamma(n \times E_t) = \operatorname{curl}_\Gamma E_t,$$

so

$$\operatorname{curl}_\Gamma E_t \in H^{-1/2}(\Gamma) \quad \Leftrightarrow \quad \operatorname{div}_\Gamma(n \times E_t) \in H^{-1/2}(\Gamma).$$

Computation of the dual norm of $n \times E_t$,

$$\|n \times E_t\|_{(H^{-1/2}(\operatorname{curl}_\Gamma, \Gamma))'} = \sup_{F_t \in H^{-1/2}(\operatorname{curl}_\Gamma, \Gamma)} \frac{|\langle n \times E_t, F_t \rangle|}{\|F_t\|_{H^{-1/2}(\operatorname{curl}_\Gamma, \Gamma)}}$$

$$= \sup_{F \in H(\operatorname{curl}, \Omega)} \frac{|\langle n \times E_t, F_t \rangle|}{\|F\|_{H(\operatorname{curl}, \Omega)}},$$

by the Riesz Representation Theorem argument reveals that the dual norm equals the $H(\operatorname{curl}, \Omega)$ norm of solution F of the Neumann problem,

$$\begin{cases} \boldsymbol{\nabla} \times (\boldsymbol{\nabla} \times F) + F = 0 & \text{in } \Omega, \\ n \times (\boldsymbol{\nabla} \times F) = n \times E_t & \text{on } \Gamma. \end{cases} \tag{5.4.42}$$

Lemma 5.12. *Let F be the solution to Neumann problem (5.4.42). Then $H = \nabla \times F \in H(\text{curl}, \Omega)$ is the solution to the corresponding Dirichlet problem,*

$$\begin{cases} \nabla \times (\nabla \times H) + H = 0 & \text{in } \Omega, \\ \qquad\qquad\qquad H_t = E_t & \text{on } \Gamma. \end{cases} \tag{5.4.43}$$

Moreover, $\|H\|_{H(\text{curl},\Omega)} = \|F\|_{H(\text{curl},\Omega)}$.

Proof. Take the curl of (5.4.42)$_1$ to learn that $H = \nabla \times F$ satisfies the same equation, and note that $n \times H_t = n \times E_t$ is equivalent to $H_t = E_t$. The equality of norms follows from the fact that F satisfies (5.4.42)$_1$. $\quad\square$

We return now to Maxwell problems discussed in Section 1.4.3. Consider the Maxwell system

$$\begin{cases} \dfrac{1}{\mu} \nabla \times E + i\omega H = 0 & \text{in } \Omega, \\ \nabla \times H - \sigma E - i\omega\epsilon E = J^{\text{imp}} & \text{in } \Omega, \\ \qquad\qquad n \times E = n \times E_0 & \text{on } \Gamma_E, \\ \qquad\qquad n \times H = n \times H_0 & \text{on } \Gamma_H. \end{cases}$$

Multiplying the second (Ampère) equation with $-i\omega$, testing it with a broken test function F, integrating over an element K, and summing up over all elements, we obtain

$$(-i\omega H, \nabla_h \times F) - ((\omega^2\epsilon - i\omega\sigma)E, F) - i\omega\langle n \times H, F\rangle_{\Gamma_h} = -i\omega(J^{\text{imp}}, F), \quad F \in H(\text{curl}, \mathcal{T}_h),$$

where

$$\langle n \times H, F\rangle_{\Gamma_h} = \sum_{K \in \mathcal{T}_h} \langle n \times H, F_K\rangle_{\partial K}.$$

As for grad-div problems, trace $n \times H = n \times H_t$ is identified as a new unknown coming from a new skeleton energy space,

$$H^{-1/2}(\text{div}, \Gamma_h) := \left\{ \hat{g} \in \prod_K H^{-1/2}(\text{div}_{\partial K}, \partial K) : \exists G \in H(\text{curl}, \Omega) \text{ such that} \right.$$

$$\left. \gamma_t^\perp(G|_K) = \hat{g} \text{ on } \partial K \right\},$$

with $\gamma_t^\perp$ denoting the rotated trace operator. We will denote the new skeleton unknown by $n \times \hat{H}_t$ where, for an element K, $\hat{H}_t$ admits a minimum energy extension $H \in H(\text{curl}, K)$ that is used to define the norm of $n \times \hat{H}_t$. Using the Faraday equation to eliminate magnetic field H, we obtain the final broken variational formulation.

$$\begin{cases} E \in H(\text{curl}, \Omega), \; n \times \hat{H}_t \in H^{-1/2}(\text{div}, \Gamma_h), \\ (\frac{1}{\mu}\nabla \times E, \nabla_h \times F) - ((\omega^2\epsilon - i\omega\sigma)E, F) - i\omega\langle n \times \hat{H}_t, F\rangle_{\Gamma_h} = -i\omega(J^{\text{imp}}, F), \\ \qquad\qquad\qquad\qquad\qquad\qquad\qquad\qquad\qquad\qquad F \in H(\text{curl}, \mathcal{T}_h), \\ \qquad\qquad\qquad\qquad\qquad\qquad E_t = E_{0,t} & \text{on } \Gamma_E, \\ \qquad\qquad\qquad\qquad\qquad n \times \hat{H}_t = n \times H_{0,t} & \text{on } \Gamma_H. \end{cases} \tag{5.4.44}$$

The well-posedness of the broken variational problem follows from Theorem 5.10.

Remark 5.13. In the same way as in the normal trace $\gamma_n v = v \cdot n$, where the normal cannot be separated from the function, the normal n in trace $n \times \hat{H}_t$ cannot be formally separated from $\hat{H}_t$ in the sense of the standard cross product as $\hat{H}_t$ is a functional rather than a function. However, by Lemma 5.12, we do have a well-defined isometric isomorphism,

$$H^{-1/2}(\mathrm{curl}_\Gamma, \Gamma) \ni \hat{H}_t \to n \times \hat{H}_t \in H^{-1/2}(\mathrm{div}_\Gamma, \Gamma) := (H^{-1/2}(\mathrm{curl}_\Gamma, \Gamma))',$$

and the cross product with n can be understood as taking the value of this isomorphism for $\hat{H}_t$. Consequently, the second BC can be replaced with $\hat{H}_t = H_{0,t}$.

In the same way, we can break the test space in any other variational formulation for the Maxwell system. For example, the broken version of the UW formulation looks as follows:

$$\begin{cases} E, H \in L^2(\Omega), \ \hat{E}_t, \hat{H}_t \in H^{-1/2}(\mathrm{curl}, \Gamma_h), \\ (\frac{1}{\mu} E, \nabla_h \times F) + \langle n \times \hat{E}_t, F_t \rangle_{\Gamma_h} + i\omega(H, F) = 0, & F \in H(\mathrm{curl}, \mathcal{T}_h), \\ (H, \nabla_h \times G) + \langle n \times \hat{H}_t, G_t \rangle_{\Gamma_h} - ((\sigma + i\omega\epsilon)E, G) = (J^{\mathrm{imp}}, G), & G \in H(\mathrm{curl}, \mathcal{T}_h), \\ \hat{E}_t = E_{0,t} & \text{on } \Gamma_E, \\ \hat{H}_t = H_{0,t} & \text{on } \Gamma_H. \end{cases}$$
$$(5.4.45)$$

Exercises

5.4.1. Write down broken versions for mixed formulations of the Maxwell system; compare Exercise 1.4.4. Exclude the impedance BCs from the discussion. (5 points)

5.5 ▪ Impedance Boundary Conditions

Implementation of impedance BCs involves additional nontrivial details. We shall first discuss the easier acoustics case and then the Maxwell equations.

5.5.1 ▪ Implementation of Impedance BC for Acoustics

Please refer to the discussion in Section 1.4.1.

Mixed formulation I and the corresponding (classical) reduced formulation I. These employ the same test space as in the case with no impedance BCs. The broken test space also remains the same—$H^1(\mathcal{T}_h)$. The relaxed continuity equation involves an additional unknown—velocity trace $\hat{u}_n \in H^{-1/2}(\Gamma_h)$,

$$i\omega(p, q) - (u, \nabla_h q) + \langle \hat{u}_n, q \rangle_{\Gamma_h} = 0, \quad q \in H^1(\mathcal{T}_h).$$

We choose to enforce the hard BC on Γ_u in a strong way, requesting

$$\hat{u}_n = u_0 \quad \text{on } \Gamma_u.$$

A similar choice for the impedance BC leads to the analogous condition on the impedance boundary,

$$\hat{u}_n = dp + u_0 \quad \text{on } \Gamma_i.$$

The condition couples two unknowns: pressure p and trace $\hat{u}_n$. A proper implementation would involve developing a variational form of the equation and discretizing the additional equation using the DPG methodology. Instead, we take a shortcut and satisfy the impedance BC *in a weak form* by building it into the relaxed continuity equation,

$$i\omega(p,q) - (u, \nabla_h q) + \langle dp, q \rangle_{\Gamma_i} + \langle \hat{u}_n, q \rangle_{\Gamma_h - \Gamma_i} = -\langle u_0, q \rangle_{\Gamma_i}, \quad q \in H^1(\mathcal{T}_h).$$

Note that, for $p, q \in H^{1/2}(\Gamma)$, duality pairing on Γ_i reduces to the L^2-product, $\langle dp, q \rangle_{\Gamma_i} = (dp, q)_{L^2(\Gamma_i)}$. For elements K adjacent to impedance boundary Γ_i, trace variable $\hat{u}_n$ is defined only on $\partial K - \Gamma_i$, in the space $\tilde{H}^{-1/2}(\partial K - \Gamma_i)$—the dual of $H^{1/2}(\partial K - \Gamma_i)$ to which the trace of test function q belongs. The corresponding norm is defined through the minimum energy extensions on K with zero boundary data on Γ_i. An appropriate regularity must be assumed on data u_0 on the impedance boundary, to make sense of coupling $\langle u_0, q \rangle_{\Gamma_i \cap \partial K}$ for each element K adjacent to the impedance boundary. For instance, assuming $u_0 \in L^2(\Gamma_i)$ is sufficient, the duality pairing on Γ_i reduces then to the L^2-product. Discretization of space $\tilde{H}^{-1/2}(\partial K - \Gamma_i)$ is done in the same way as $H^{-1/2}(\partial K - \Gamma_i)$—with traces of $H(\mathrm{div})$-conforming elements.

Mixed formulation II and the corresponding reduced formulation II. Both formulations involve relaxation of the momentum equation and employ a modified broken test space,

$$V(\mathcal{T}_h) := \left\{ v = \left\{ v_K \right\} \in \prod_K H(\mathrm{div}, K) \ : \ v_K|_{\Gamma_i} \in L^2(\Gamma_i \cap \partial K) \right\}.$$

The relaxed momentum equation looks as follows:

$$i\omega(u,v) - (p, \mathrm{div}_h v) + \langle \hat{p}, v_n \rangle_{\Gamma_h - \Gamma_i} + \langle d^{-1} u_n, v_n \rangle_{\Gamma_i} = \langle d^{-1} u_0, v_n \rangle_{\Gamma_i}, \quad v \in V(\mathcal{T}_h).$$

The solution u comes from the modified energy space V. The modification of the test space includes employing a stronger test inner product for elements K adjacent to impedance boundary Γ_i,

$$(v, \delta v)_{V(K)} := (v, \delta v)_{L^2(K)} + (\mathrm{div}\, v, \mathrm{div}\, \delta v)_{L^2(K)} + (v_n, \delta v_n)_{L^2(\Gamma_i \cap \partial K)}. \tag{5.5.46}$$

Trace $\hat{p}$ lives on $\partial \Gamma_h - \Gamma_i$ and elementwise belongs to space $H^{1/2}(\partial K - \Gamma_i)$. Its norm is defined through the minimum energy extension problem with an impedance BC on $\Gamma_i \cap \partial K$,

$$\begin{cases} p \in H^1(K), \\ \qquad\qquad p = \hat{p} \quad \text{on } \partial K - \Gamma_i, \\ -p + \dfrac{\partial p}{\partial n} = 0 \quad \text{on } \partial K \cap \Gamma_i, \\ -\Delta p + p = 0 \quad \text{in } K. \end{cases} \tag{5.5.47}$$

The modified minimum energy extension norm

$$\|\hat{p}\|_E^2 := \|p\|_{H^1(K)}^2 + \|p\|_{L^2(\partial K \cap \Gamma_i)}^2$$

and the "enriched" test norm corresponding to inner product (5.5.46) are in duality pairing; compare Exercise 5.5.1. Boundary-value problem (5.5.47) implies that $\hat{p} \in H^{1/2}(\partial K - \Gamma_i)$. This can also be seen by noticing that the restriction of normal trace v_n to $\partial K - \Gamma_i$, by assumption, can be extended by an L^2-function to the whole element boundary ∂K. But, in turn, an L^2-function on $\partial K \cap \Gamma_i$ admits a zero extension to the whole boundary. Subtracting it from v_n, we realize that v_n *admits a zero extension* to the whole boundary and, therefore, lives in $\tilde{H}^{-1/2}(\partial K - \Gamma_i)$. This implies that trace $\hat{p}$ lives in the dual space $H^{1/2}(\partial K - \Gamma_i)$. This hair-splitting exercise in energy spaces ensures us that trace $\hat{p}$ can be discretized conformingly with simple restrictions of H^1-conforming elements to $\Gamma_h - \Gamma_i$.

Ultraweak formulation. The relaxed continuity and momentum equations look as follows:

$$i\omega(p, q) - (u, \boldsymbol{\nabla}_h q) + \langle \hat{u}_n, q\rangle_{\Gamma_h} = 0, \qquad q \in H^1(\mathcal{T}_h),$$

$$i\omega(u, v) - (p, \operatorname{div}_h v) + \langle \hat{p}, v_n\rangle_{\Gamma_h} = 0, \qquad v \in H(\operatorname{div}, \mathcal{T}_h).$$

We have two choices on impedance boundary Γ_i: replace $\hat{u}_n$ with $\hat{p}$ or, vice versa, $\hat{p}$ with $\hat{u}_n$. The first one is easier as we do not have to modify the energy setting. Then we obtain

$$\begin{cases} p \in L^2(\Omega), u \in L^2(\Omega)^N, \\ \hat{p} \in H^{1/2}(\Gamma_h), \hat{p} = p_0 \text{ on } \Gamma_p, \\ \hat{u}_n \in \tilde{H}^{-1/2}(\Gamma_h - \Gamma_i), \hat{u}_n = u_0 \text{ on } \Gamma_u, \\ i\omega(p, q) - (u, \boldsymbol{\nabla}_h q) + \langle \hat{u}_n, q\rangle_{\Gamma_h - \Gamma_i} + (d\hat{p}, q)_{L^2(\Gamma_i)} = -(u_0, q)_{L^2(\Gamma_i)}, q \in H^1(\mathcal{T}_h), \\ i\omega(u, v) - (p, \operatorname{div}_h v) + \langle \hat{p}, v_n\rangle_{\Gamma_h} = 0, \qquad v \in H(\operatorname{div}, \mathcal{T}_h). \end{cases}$$

5.5.2 ▪ Implementation of Impedance BC for Maxwell Equations

Please refer to the discussion in Section 1.4.3.

Reduced formulation in terms of the electric field. We begin by testing the Ampère equation with a broken test function $F \in H(\operatorname{curl}, \mathcal{T}_h)$, and introducing a new unknown—trace $\hat{H}_t$,

$$-i\omega(H, \boldsymbol{\nabla}_h \times F) - ((\omega^2\epsilon - i\omega\sigma)E, F) - i\omega\langle n \times \hat{H}_t, F\rangle_{\Gamma_h} = -i\omega(J^{\mathrm{imp}}, F), \quad F \in H(\operatorname{curl}, \mathcal{T}_h).$$

Formally building the impedance BC into the formulation, we obtain

$$-i\omega(H, \boldsymbol{\nabla}_h \times F) - ((\omega^2\epsilon - i\omega\sigma)E, F) - i\omega\langle n \times \hat{H}_t, F\rangle_{\Gamma_h - \Gamma_i} - i\omega\langle d\hat{E}_t, F\rangle_{\Gamma_i}$$
$$= -i\omega(J^{\mathrm{imp}}, F) + i\omega\langle J_S^{\mathrm{imp}}, F\rangle_{\Gamma_i}, \quad F \in H(\operatorname{curl}, \mathcal{T}_h).$$

To rigorously define the terms on the impedance boundary, we have to introduce a new broken test space,

$$V(\mathcal{T}_h) = \{F = \{F_K\} : F_K \in H(\operatorname{curl}, K) : F_{K,t}|_{\partial K \cap \Gamma_i} \in L^2(\partial K \cap \Gamma_i)\},$$

and a new space for traces,

$$\hat{U} := \left\{\{\hat{H}_{K,t}\} \in \prod_K H^{-1/2}(\operatorname{curl}, \partial K) : \exists H \in H(\operatorname{curl}, \Omega), H|_{\Gamma_i} \in L^2(\Gamma_i)\right.$$
$$\left. \text{such that } \gamma_t H|_K = \hat{H}_{K,t}\right\}.$$

The unknown trace $\hat{H}_t$ comes from a space $\hat{U}(\Gamma_h - \Gamma_i)$ consisting of *restrictions* of functions from $\hat{U}$ to $\Gamma_h - \Gamma_i$, i.e., elementwise to $\partial K - \Gamma_i$.

What are the practical modifications in the implementation? The first one deals with the modified broken test inner product that now includes the $L^2(\Gamma_i)$ term,

$$(F, \delta F)_{V(K)} = (F, \delta F)_{H(\operatorname{curl}, K)} + (F_t, \delta F_t)_{L^2(\partial K \cap \Gamma_i)}. \tag{5.5.48}$$

The second one deals with the modified energy extension norm in which the new trace is measured,

$$\|\hat{H}_t\|^2 = \|H\|^2_{H(\operatorname{curl}, K)} + \|H_t\|^2_{L^2(\partial K \cap \Gamma_i)},$$

where H solves the following problem involving an impedance BC on $\partial K \cap \Gamma_i$:

$$\begin{cases} H \in H(\text{curl}, K), H_t|_{\partial K \cap \Gamma_i} \in L^2(\partial K \cap \Gamma_i), \\ \qquad\qquad H_t = \hat{H}_t \quad \text{on } \partial K - \Gamma_i, \\ n \times (\nabla \times H) + H_t = 0 \quad \text{on } \partial K \cap \Gamma_i, \\ \nabla \times (\nabla \times H) + H = 0 \quad \text{in } K; \end{cases} \qquad (5.5.49)$$

see Exercise 5.5.2.

The ultimate DPG variational formulation looks as follows:

$$\begin{cases} E \in H(\text{curl}, \Omega), E_t|_{\Gamma_i} \in L^2(\Gamma_i), E_t = E_{0,t} \text{ on } \Gamma_E, \\ \hat{H}_t \in \hat{U}(\Gamma_h - \Gamma_i), \hat{H}_t = H_{0,t} \text{ on } \Gamma_H, \\ (\mu^{-1}\nabla \times E, \nabla_h \times F) - ((\omega^2 \epsilon - i\omega\sigma)E, F) - i\omega\langle n \times \hat{H}_t, F \rangle_{\Gamma_h - \Gamma_i} + i\omega(dE_t, F_t)_{L^2(\Gamma_i)} \\ \qquad\qquad = i\omega(J^{\text{imp}}, F) + i\omega(J_S^{\text{imp}}, F)_{L^2(\Gamma_i)}, \qquad F \in V(\mathcal{T}_h), \end{cases}$$

with the assumption that $J_S^{\text{imp}} \in L^2(\Gamma_i)$. We summarize that the extra regularity assumptions on traces do not affect the standard discretization with traces of $H(\text{curl})$-conforming elements, but they do require a modified test inner product for elements adjacent to the impedance boundary.

Ultraweak variational formulation. For reference, the broken variational formulation with impedance BC condition is given by

$$\begin{cases} E, H \in L^2(\Omega)^3, \\ \hat{E}_t \in \hat{U}, \hat{E}_t = E_{0,t} \text{ on } \Gamma_E, \\ \hat{H}_t \in \hat{U}(\Gamma_h - \Gamma_i), \hat{H}_t = H_{0,t} \text{ on } \Gamma_H, \\ (\mu^{-1}E, \nabla_h \times F) + \langle n \times \hat{E}_t, F_t \rangle_{\Gamma_h} + i\omega(H, F) = 0, \quad F \in H(\text{curl}, \mathcal{T}_h), \\ (H, \nabla_h \times G) + \langle n \times \hat{H}_t, G_t \rangle_{\Gamma_h - \Gamma_i} + (d\hat{E}_t, G_t)_{L^2(\Gamma_i)} - ((\sigma + i\omega\epsilon)E, G) \\ \qquad\qquad = (J^{\text{imp}}, G) + (J_S^{\text{imp}}, F)_{L^2(\Gamma_i)} \quad G \in V(\mathcal{T}_h). \end{cases}$$

Contrary to the UW formulation for acoustics that required no changes in the test norms, the UW formulation for Maxwell's equations does require a small upgrade of the test space and norm for the relaxed Ampère equation.

Exercises

5.5.1. Prove that the test norm corresponding to inner product (5.5.46) and the minimum energy extension norm defined by problem (5.5.47) are in duality pairing. *Hint:* Follow reasoning from Lemma 5.8. (5 points)

5.5.2. Prove that the test norm corresponding to inner product (5.5.48) and the minimum energy extension norm defined by problem (5.5.49) are in duality pairing. *Hint:* Follow reasoning from Lemma 5.12. (5 points)

5.6 ▪ Construction of Fortin Operators for DPG Problems

Recall the abstract conditions for the Fortin operator in the context of the DPG method:

$$\Pi \ : \ V(\mathcal{T}_h) \ni v \to \Pi v \in V_h(\mathcal{T}_h) \,,$$

$$\|\Pi v\|_{V(\mathcal{T}_h)} \le C_F \|v\|_{V(\mathcal{T}_h)} \,,$$

$$b(u_h, v - \Pi v) + \langle \hat{u}_h, v - \Pi v \rangle_{\Gamma_h} = 0 \quad \forall u_h \in U_h, \ \hat{u}_h \in \hat{U}_h \,.$$

Construction of Fortin operators for conforming test spaces is challenging. The value of the operator, Πv, has to be in the (conforming) discrete test space, which suggests the use of techniques applied in the construction of interpolation operators: taking values at vertices, edge and face averages, etc. However, the Fortin operator has to be defined on the *whole energy space*, and these operations are not well-defined for general members of such spaces.

With broken test spaces, the global conformity is not an issue, and we can settle for a local construction of the Fortin operator:

$$\Pi \ : \ V(K) \ni v \to \Pi v \in V_h(K) \,,$$

$$\|\Pi v\|_{V(K)} \le C_F \|v\|_{V(K)} \,, \tag{5.6.50}$$

$$b_K(u_h, v - \Pi v) + \langle \hat{u}_h, v - \Pi v \rangle_{\partial K} = 0 \quad \forall u_h \in U_h, \ \hat{u}_h \in \hat{U}_h \,.$$

Clearly, satisfaction of the local conditions implies immediately satisfaction of the global conditions as well. The main point in the construction of the Fortin operator is to use operations that are well-defined and continuous on the whole energy space. The finite-dimensionality of its range and the closedness of its null space then automatically imply the continuity of the operator; see Exercise 5.6.3. We also want the continuity constant to be at least independent of element size h and, possibly, independent of polynomial order p. As the Fortin constant enters the ultimate stability constant for the practical DPG method, we also want it to be as small as possible.

Construction of the Fortin operator involves the original bilinear form and the skeleton term resulting from breaking the test space and, therefore, is problem dependent. However, if we restrict ourselves to standard test spaces, H^1, $H(\text{curl})$, $H(\text{div})$ (with standard norms), and make a simplifying assumption about the material data to be elementwise constant, one can strive for constructing general Fortin operators that will serve all problems satisfying the simplifying assumptions. This was done in [50, 16]. Below, we follow [39] generalizing ideas from [61].

We will restrict ourselves to affine tetrahedral elements.

The motivation for the construction comes from the UW variational formulation for two model problems. The first one is the classical diffusion-convection-reaction problem:

$$\begin{cases} -\operatorname{div}\sigma + cu = f & \text{in } \Omega \,, \\ a^{-1}\sigma - \nabla u + a^{-1}bu = 0 & \text{in } \Omega \,, \\ u = u_0 & \text{on } \Gamma_u \,, \\ \sigma \cdot n = \sigma_0 & \text{on } \Gamma_\sigma \,. \end{cases}$$

An element K contribution to the bilinear form in the UW variational formulation is

$$b_K((\sigma, u, \hat{\sigma} \cdot n, \hat{u}), (\tau, v))$$
$$= (\sigma, \nabla v + a^{-1}\tau)_K + (u, cv + \operatorname{div}\tau + (a^{-1}b)\cdot\tau)_K - \langle \hat{\sigma}_n, v \rangle_{\partial K} - \langle \hat{u}, \tau \cdot n \rangle_{\partial K} \,,$$

where, consistent with the logic of using the exact sequence spaces for discretization, we have

$$u \in \mathcal{P}^{p-1}(K), \sigma \in \mathcal{P}^{p-1}(K)^3\,,$$
$$\hat{u} \in \gamma(\mathcal{P}^p(K)) =: \mathcal{P}_c^p(\partial K)\,,$$
$$\hat{\sigma}_n \in \gamma_n(\mathcal{RT}^p(K)) =: \mathcal{P}_d^{p-1}(\partial K)\,.$$

After integration by parts,

$$b_K((\sigma, u, \hat{\sigma}_n, \hat{u}), (\tau, v))$$
$$= (a^{-1}\sigma - \nabla u + a^{-1}bu, \tau)_K + (-\operatorname{div}\sigma + cu, v)_K + \langle\sigma_n - \hat{\sigma}_n, v\rangle_{\partial K} + \langle u - \hat{u}, \tau_n\rangle_{\partial K}\,.$$

This leads to the following orthogonality requirements for the Fortin operators:

$$\begin{aligned}
(\psi, \Pi^{\mathrm{grad}}v - v)_K &= 0 &&\forall\psi \in \mathcal{P}^{p-1}(K)\,, \\
\langle\phi, \Pi^{\mathrm{grad}}v - v\rangle_{\partial K} &= 0 &&\forall\phi \in \mathcal{P}_d^{p-1}(\partial K)\,,
\end{aligned} \tag{5.6.51}$$

$$\begin{aligned}
(\psi, \Pi^{\mathrm{div}}\tau - \tau)_K &= 0 &&\forall\psi \in \mathcal{P}^{p-1}(K)^3\,, \\
\langle\phi, (\Pi^{\mathrm{div}}\tau - \tau)\cdot n\rangle_{\partial K} &= 0 &&\forall\phi \in \mathcal{P}_c^p(\partial K)\,.
\end{aligned} \tag{5.6.52}$$

Our second example deals with the UW formulation for the 3D Maxwell equations,

$$\begin{cases}
E, H \in L^2(\Omega),\ \hat{E}_t, \hat{H}_t \in H^{-1/2}(\operatorname{curl}_\Gamma, \Gamma)\,, \\
(\frac{1}{\mu}E, \nabla_h \times F) + \langle n \times \hat{E}_t, F_t\rangle_{\Gamma_h} + i\omega(H, F) = 0, & F \in H(\operatorname{curl}, \mathcal{T}_h)\,, \\
(H, \nabla_h \times G) + \langle n \times \hat{H}_t, G_t\rangle_{\Gamma_h} - ((\sigma + i\omega\epsilon)E, G) = (J^{\mathrm{imp}}, G), & G \in H(\operatorname{curl}, \mathcal{T}_h)\,, \\
\hat{E}_t = E_{0,t} & \text{on } \Gamma_E\,, \\
\hat{H}_t = H_{0,t} & \text{on } \Gamma_H\,.
\end{cases}$$

Recalling that approximate $E, H \in \mathcal{P}^{p-1}(K)^3$ and approximate $\hat{E}_t, \hat{H}_t$ belong to the tangential trace of $\mathcal{N}^p(K)$, we arrive at the orthogonality conditions for the Fortin operator:

$$\begin{aligned}
(\psi, \Pi^{\mathrm{curl}}F - F)_K &= 0, &&\psi \in \mathcal{P}^{p-1}(K)^3\,, \\
\langle n \times \phi, \Pi^{\mathrm{curl}}F - F\rangle_{\partial K} &= 0, &&\phi \in \gamma_t\mathcal{N}^p(K)\,,
\end{aligned} \tag{5.6.53}$$

where $\gamma_t\mathcal{N}^p(K)$ denotes the image of tangential trace operator of $\mathcal{N}^p(K)$.

5.6.1 ▪ Auxiliary Results

We will need a few fundamental results on polynomial spaces defined on the master tetrahedron K. The first four lemmas deal with bubble spaces.

Lemma 5.14. *Let $\mathcal{P}_0^{p+3}(K)$ denote the subspace of $\mathcal{P}^{p+3}(K)$ of H^1 bubbles on element K. Let $u \in \mathcal{P}_0^{p+3}(K)$, and*

$$(\psi, u)_K = 0 \quad \forall\psi \in \mathcal{P}^{p-1}(K)\,.$$

Then $u = 0$ and, consequently,

$$\inf_{u \in \mathcal{P}_0^{p+3}(K)} \sup_{\psi \in \mathcal{P}^{p-1}(K)} \frac{|(\psi, u)_K|}{\|\psi\|\,\|u\|} = \beta = \beta(p) > 0\,.$$

As spaces $\mathcal{P}_0^{p+3}(K)$ and $\mathcal{P}^{p-1}(K)$ are of equal dimension (compare Table 3.1), the order of spaces in the inf-sup condition can be reversed,

$$\inf_{\psi \in \mathcal{P}^{p-1}(K)} \sup_{u \in \mathcal{P}_0^{p+3}(K)} \frac{|(\psi, u)_K|}{\|u\|\,\|\psi\|} = \beta > 0 \,.$$

Proof. Function u must be of the form

$$u = \lambda_0 \ldots \lambda_3 \, v \,,$$

where λ_i, $i = 0, \ldots, 3$, are affine coordinates, and $v \in \mathcal{P}^{p-1}(K)$. Choosing $\psi = v$ gives

$$(\psi, u)_K = \int_K \lambda_0 \ldots \lambda_3 \, v^2 = 0 \quad \Rightarrow \quad v = 0 \quad \Rightarrow \quad u = 0 \,.$$

The result implies that the supremum

$$\sup_{\psi \in \mathcal{P}^{p-1}(K)} \frac{|(\psi, u)_K|}{\|\psi\|}$$

defines a norm on u, and the inf-sup condition then follows from the equivalence of norms in a finite-dimensional space.   □

The following result holds for both master triangle ($d = 2$) and master tetrahedron ($d = 3$), and it can be found in [62].

Lemma 5.15. Let $\mathcal{RT}_0^{p+1}(K)$ denote the subspace of $\mathcal{RT}^{p+1}(K)$ of $H(\mathrm{div})$ bubbles on element K. Let $\tau \in \mathcal{RT}_0^{p+1}(K)$, and

$$(\psi, \tau)_K = 0 \quad \forall \psi \in \mathcal{P}^{p-1}(K)^d \,.$$

Then $\tau = 0$ and, consequently,

$$\inf_{\tau \in \mathcal{RT}_0^{p+1}(K)} \sup_{\psi \in \mathcal{P}^{p-1}(K)^d} \frac{|(\psi, \tau)_K|}{\|\psi\|\,\|\tau\|} = \beta = \beta(p) > 0 \,.$$

As spaces $\mathcal{RT}_0^{p+1}(K)$ and $\mathcal{P}^{p-1}(K)^d$ are of equal dimension (compare Table 3.1), the order of spaces in the inf-sup condition can be reversed,

$$\inf_{\psi \in \mathcal{P}^{p-1}(K)^d} \sup_{\tau \in \mathcal{RT}_0^{p+1}(K)} \frac{|(\psi, \tau)_K|}{\|\tau\|\,\|\psi\|} = \beta > 0 \,.$$

Proof. Take $\psi = \boldsymbol{\nabla} \mathrm{div} \, \tau$. Integration by parts reveals that $\mathrm{div} \, \tau = 0$. This implies that τ is the curl of an element of Nédélec space $\mathcal{N}^p(K)$ and, in particular, it must be a polynomial of order p, i.e., $\tau \in \mathcal{P}^p(K)^d$. As τ satisfies the homogeneous normal BC, there must exist $\psi_i \in \mathcal{P}^{p-1}(K)$ such that

$$\tau_i = \xi_i \psi_i \,.$$

Testing with such a ψ gives

$$\int_K \tau \psi = \int_K \sum_i \xi_i |\psi_i|^2 = 0 \quad \Rightarrow \quad \psi = 0 \quad \Rightarrow \quad \tau = 0 \,.$$

The result implies that the supremum

$$\sup_{\psi \in \mathcal{P}^{p-1}(K)^d} \frac{|(\psi, \tau)_K|}{\|\psi\|}$$

defines a norm on τ, and the inf-sup condition then follows from the equivalence of norms in a finite-dimensional space. $\qquad \square$

Lemma 5.16. *Let $\mathcal{N}_0^{p+2}(K)$ denote the subspace of $\mathcal{N}^{p+2}(K)$ of $H(\mathrm{curl})$ bubbles defined on tetrahedron K. Let $F \in \mathcal{N}_0^{p+2}(K)$, and*

$$(\psi, F)_K = 0 \quad \forall \psi \in \mathcal{P}^{p-1}(K)^3 \,.$$

Then $F = 0$ and, consequently,

$$\inf_{F \in \mathcal{N}_0^{p+2}(K)} \sup_{\psi \in \mathcal{P}^{p-1}(K)^3} \frac{|(\psi, F)_K|}{\|\psi\| \, \|F\|} = \beta = \beta(p) > 0 \,.$$

As spaces $\mathcal{N}_0^{p+2}(K)$ and $\mathcal{P}^{p-1}(K)^3$ are of equal dimension (compare Table 3.1), the order of spaces in the inf-sup condition can be reversed,

$$\inf_{\psi \in \mathcal{P}^{p-1}(K)^3} \sup_{F \in \mathcal{N}_0^{p+2}(K)} \frac{|(\psi, F)_K|}{\|\psi\| \, \|F\|} = \beta > 0 \,.$$

Proof. Let $F \in \mathcal{N}_0^{p+2}(K)$. Let $\psi \in \mathcal{P}^p(K)^3$. Then

$$(\psi, \nabla \times F)_K = (\nabla \times \psi, F)_K = 0 \,.$$

As the curl operator sets $H(\mathrm{curl})$ bubbles into $H(\mathrm{div})$ bubbles, Lemma 5.15 proves that $\nabla \times F = 0$ and, in particular, $F \in \mathcal{P}^{p+1}(K)^3$. Any $H(\mathrm{curl})$ bubble on the master tetrahedron must be of the form

$$F = (\phi_1 \, \xi_2 \xi_3, \phi_2 \, \xi_1 \xi_3, \phi_3 \, \xi_1 \xi_2)$$

with some scalar factors ϕ_i. As F is of order $p + 1$, ϕ_i must be of order $p - 1$. Selecting $\psi = (\phi_1, \phi_2, \phi_3)$, we conclude that $F = 0$. The rest of the reasoning is the same as in the proof of Lemma 5.15. $\qquad \square$

In order to cope with boundary terms, we will also need a 2D equivalent of Lemma 5.16.

Lemma 5.17. *Let $\mathcal{N}_0^{p+1}(K)$ denote the subspace of $\mathcal{N}^{p+1}(K)$ of $H(\mathrm{curl})$ bubbles on the master triangle K. Let $F \in \mathcal{N}_0^{p+1}(K)$, and*

$$(\psi, F)_K = 0 \quad \forall \psi \in \mathcal{P}^{p-1}(K)^2 \,.$$

Then $F = 0$ and, consequently,

$$\inf_{F \in \mathcal{N}_0^{p+1}(K)} \sup_{\psi \in \mathcal{P}^{p-1}(K)^2} \frac{|(\psi, F)_K|}{\|\psi\| \, \|F\|} = \beta = \beta(p) > 0 \,.$$

As spaces $\mathcal{N}_0^{p+1}(K)$ and $\mathcal{P}^{p-1}(K)^2$ are of equal dimension (compare Table 3.5), the order of spaces in the inf-sup condition can be reversed,

$$\inf_{\psi \in \mathcal{P}^{p-1}(K)^2} \sup_{F \in \mathcal{N}_0^{p+2}(K)} \frac{|(\psi, F)_K|}{\|\psi\| \, \|F\|} = \beta > 0 \,.$$

Proof. The result follows directly from the 2D version of Lemma 5.15 and the relation between the two 2D exact sequences. See also Exercise 5.6.5.   □

The next three lemmas deal with polynomial spaces satisfying the orthogonality constraints necessary for Fortin operators. We will slightly upgrade the orthogonality assumptions $(5.6.53)_2$, replacing them with

$$(\psi, \Pi^{\text{curl}} F - F)_K = 0 \quad \forall \psi \in \mathcal{P}^{p-1}(K)^3 \,,$$
$$\langle n \times \phi, \Pi^{\text{curl}} F - F \rangle_{\partial K} = 0 \quad \forall \phi \in \gamma_t(\mathcal{P}^p(K)^3) \,. \tag{5.6.54}$$

Lemma 5.18. *Let $F \in H(\text{curl}, K)$ satisfy the constraints*

$$(\psi, F)_K = 0 \quad \forall \psi \in \mathcal{P}^{p-1}(K)^3 \,,$$
$$\langle n \times \phi, F \rangle_{\partial K} = 0 \quad \forall \phi \in \mathcal{P}^p(K)^3 \,. \tag{5.6.55}$$

Then $\text{curl} F$ *satisfies the constraint*

$$(\chi, \text{curl} F)_K = 0 \quad \forall \chi \in \mathcal{P}^p(K)^3 \,, \tag{5.6.56}$$

which, in turn, implies

$$\langle \eta, \text{curl} F \cdot n \rangle_{\partial K} = 0 \quad \forall \eta \in \mathcal{P}^{p+1}(K) \,. \tag{5.6.57}$$

Conversely, let $F \in H(\text{curl}, K)$ *satisfy* (5.6.56). *Then there exists* $u \in \mathcal{P}^{p+2}(K)$ *such that*

$$(\psi, F + \nabla u)_K = 0 \quad \forall \psi \in \mathcal{P}^{p-1}(K)^3 \text{ and}$$
$$\langle n \times \phi, F + \nabla u \rangle_{\partial K} = 0 \quad \forall \phi \in \mathcal{P}^p(K)^3 \,. \tag{5.6.58}$$

Proof. Taking $\psi = \text{curl}\, \chi$ in $(5.6.55)_1$ and utilizing $(5.6.55)_2$ gives (5.6.56). Use $\chi = \nabla \eta$ in (5.6.56) to obtain (5.6.57).

Let $F \in H(\text{curl}, K)$ now satisfy (5.6.56). It is sufficient to show $(5.6.58)_1$, i.e., that the variational problem,

$$\begin{cases} u \in \mathcal{P}^{p+2}(K) \,, \\ (\nabla u, \delta \psi)_K = -(F, \delta \psi)_K, \quad \delta \psi \in \mathcal{P}^{p-1}(K)^3 \,, \end{cases} \tag{5.6.59}$$

has a solution u. The second property follows from the first one with $\psi = \nabla \times \phi$ and (5.6.56). We begin by considering the null space of the conjugate operator,

$$\{\psi \in \mathcal{P}^{p-1}(K)^3 \,:\, (\nabla \delta u, \psi)_K = 0 \quad \forall \delta u \in \mathcal{P}^{p+2}(K)\} \,.$$

We claim that the constraint for ψ is equivalent to $\psi = \text{curl}\, \zeta$, where $\zeta \in \mathcal{P}^p(K)^3$ with a zero tangential trace. Sufficiency follows from integration by parts. To show necessity, we test first with $\delta u \in \mathcal{P}_0^{p+2}(K)$ to obtain

$$(\underbrace{\text{div}\, \psi}_{\in \mathcal{P}^{p-2}(K)}, \delta u)_K = 0 \,.$$

Taking $\delta u = \text{div}\, \psi\, \lambda_0 \ldots \lambda_3$, where λ_i, $i = 0, \ldots, 3$, are affine coordinates, we conclude that $\text{div}\, \psi = 0$. Testing next with a general δu, we obtain

$$0 = (\nabla \delta u, \psi)_K = \langle \delta u, \psi \cdot n \rangle_{\partial K} \,.$$

Taking $\delta u = (\psi \cdot n)\lambda_i \lambda_j \lambda_k$ on each $[ijk]$ face, we conclude that $\psi \cdot n = 0$ on ∂K. Consequently, there exists a vector potential $\zeta \in \mathcal{P}^p(K)^3$ with zero tangential trace such that $\psi = \operatorname{curl} \zeta$; compare Remark 3.5.

To finish the proof, we need to notice that condition (5.6.56) on F implies that the right-hand side of variational problem (5.6.59) is orthogonal to the null space of the transpose operator. Indeed,

$$(F, \operatorname{curl} \zeta)_K = (\operatorname{curl} F, \zeta)_K = 0 \qquad \forall \zeta \in \mathcal{P}^p(K)^3 \text{ with a zero tangential trace.} \qquad \Box$$

Lemma 5.19. *Let $\tau \in H(\operatorname{div}, K)$ satisfy the constraints*

$$
\begin{aligned}
(\psi, \tau)_K &= 0 \quad \forall \psi \in \mathcal{P}^{p-1}(K)^3\,, \\
\langle \phi, \tau \cdot n \rangle_{\partial K} &= 0 \quad \forall \phi \in \mathcal{P}^p(K)\,.
\end{aligned}
\tag{5.6.60}
$$

Then $\operatorname{div} \tau$ satisfies the constraint

$$(\chi, \operatorname{div} \tau)_K = 0 \quad \forall \chi \in \mathcal{P}^p(K)\,. \tag{5.6.61}$$

Conversely, let $\tau \in H(\operatorname{div}, K)$ satisfy (5.6.61). Then there exists $F \in \mathcal{N}^{p+1}(K)$ such that

$$
\begin{aligned}
(\psi, \tau + \operatorname{curl} F)_K &= 0 \quad \forall \psi \in \mathcal{P}^{p-1}(K)^3 \text{ and} \\
\langle \phi, (\tau + \operatorname{curl} F) \cdot n \rangle_{\partial K} &= 0 \quad \forall \phi \in \mathcal{P}^p(K)\,.
\end{aligned}
\tag{5.6.62}
$$

Proof. Taking $\psi = \nabla \chi$ in $(5.6.60)_1$ and utilizing $(5.6.60)_2$ gives (5.6.61).

Now let τ satisfy (5.6.61). In the same way as in the proof of Lemma 5.18, we will prove that the variational problem,

$$
\begin{cases}
F \in \mathcal{N}^{p+1}(K)\,, \\
(\operatorname{curl} F, \delta\psi)_K = -(\tau, \delta\psi)_K\,, \qquad \delta\psi \in \mathcal{P}^{p-1}(K)^3\,,
\end{cases}
\tag{5.6.63}
$$

has a solution F. The null space of the transpose operator is equal to

$$\{\psi \in \mathcal{P}^{p-1}(K)^3 \,:\, (\operatorname{curl} \delta F, \psi)_K = 0 \quad \forall \delta F \in \mathcal{N}^{p+1}(K)\}\,.$$

We claim that ψ satisfies the constraint iff $\psi = \nabla u$, $u \in \mathcal{P}_0^p(K)$. The sufficiency follows from integration by parts. In order to prove necessity, we first test with $\delta F_0 \in \mathcal{N}^{p+1}(K)$ with zero tangential trace. We obtain

$$(\delta F_0, \underbrace{\operatorname{curl} \psi}_{\in \mathcal{P}^{p-2}(K)^3})_K = 0$$

and, by Lemma 5.16, $\operatorname{curl} \psi = 0$. Testing next with a general F and using Lemma 5.17, we conclude that $\gamma_t \psi = 0$ on ∂K. Consequently, there exists a $u \in \mathcal{P}_0^p(K)$ such that $\psi = \nabla u$.

It is now sufficient to notice that the right-hand side in variational problem (5.6.63) is orthogonal to the null space of the transpose operator,

$$-(\tau, \nabla u)_K = (\operatorname{div} \tau, u)_K = 0 \qquad \forall u \in \mathcal{P}_0^p(K)\,.$$

Finally, property $(5.6.62)_2$ follows from testing in $(5.6.62)_1$ with $\psi = \nabla \phi$, $\phi \in \mathcal{P}^p(K)$, integration by parts, and (5.6.61). $\Box$

In the following lemma, we upgrade slightly condition $(5.6.51)_2$.

Lemma 5.20. *Let $u \in H^1(K)$ satisfy the constraints*

$$
\begin{aligned}
(\psi, u)_K &= 0 \quad \forall \psi \in \mathcal{P}^{p-1}(K)\,, \\
\langle \phi \cdot n, u \rangle_{\partial K} &= 0 \quad \forall \phi \in \mathcal{P}^p(K)^3\,.
\end{aligned}
\tag{5.6.64}
$$

Then ∇u satisfies the constraint

$$
(\chi, \nabla u)_K = 0 \quad \forall \chi \in \mathcal{P}^p(K)^3\,,
\tag{5.6.65}
$$

which, in turn, implies

$$
\langle n \times \eta, \nabla u \rangle_{\partial K} = 0 \quad \forall \eta \in \mathcal{P}^{p+1}(K)^3\,.
\tag{5.6.66}
$$

Conversely, let $u \in H^1(K)$ satisfy (5.6.65). Then there exists a constant c such that

$$
\begin{aligned}
(\psi, u + c)_K &= 0 \quad \forall \psi \in \mathcal{P}^{p-1}(K) \text{ and} \\
\langle \phi \cdot n, u + c \rangle_{\partial K} &= 0 \quad \forall \phi \in \mathcal{P}^p(K)^3\,.
\end{aligned}
\tag{5.6.67}
$$

Proof. See Exercise 5.6.1.   □

5.6.2 ▪ Π^{div} Fortin Operator.

We begin with the construction of the Π^{div} Fortin operator. The idea is to first construct operator $\hat{\Pi}^{\mathrm{div}}$ on master tetrahedron $\hat{K}$, and then use the $H(\mathrm{div})$ pullback map T to extend it to an arbitrary affine element K,

$$
\Pi^{\mathrm{div}} \tau := T^{-1} \hat{\Pi}^{\mathrm{div}} T \tau\,.
$$

Similar to the interpolation error estimates, the scaling properties of pullback maps imply that we should have the commuting diagram

$$
\begin{array}{ccc}
H(\mathrm{div}, K) & \xrightarrow{\mathrm{div}} & L^2(K) \\
\Pi^{\mathrm{div}} \downarrow & & P \downarrow \\
V^{p+1} & \xrightarrow{\mathrm{div}} & Y^p\,,
\end{array}
\tag{5.6.68}
$$

where V^{p+1} is the enriched $H(\mathrm{div})$ test space, $Y^p = \mathrm{div}\, V^{p+1}$, and P is a Fortin operator for the L^2-space. In other words, divergence of $\Pi^{\mathrm{div}} \tau$ should only depend upon the divergence of function τ.

We will show now how the required orthogonality properties and the commutativity of Fortin operators lead in a logical way to optimal definitions of the four Fortin operators.

The required orthogonality properties for the Π^{div} operator,

$$
\begin{aligned}
(\Pi^{\mathrm{div}} \tau - \tau, \psi)_K &= 0 \quad \forall \psi \in \mathcal{P}^{p-1}(K)^3\,, \\
\langle (\Pi^{\mathrm{div}} \tau - \tau) \cdot n, \phi \rangle_{\partial K} &= 0 \quad \forall \phi \in \mathcal{P}^p(K)\,,
\end{aligned}
\tag{5.6.69}
$$

Lemma 5.19, and commuting property (5.6.68) imply orthogonality properties for operator P : $Y \to Y^p$,

$$
(Py - y, \chi) \quad \forall \chi \in \mathcal{P}^p(K)\,.
\tag{5.6.70}
$$

This leads to the definition of P to be the L^2-projection onto Y^p,

$$
\|y^p - y\| \to \min_{y^p \in Y^p}\,,
\tag{5.6.71}
$$

and the minimum assumption on space Y^p:

$$\mathcal{P}^p \subset Y^p \,.$$

Once we have defined $y^p = \operatorname{div} \tau^{p+1}$, we proceed with a second, constrained minimization problem to complete the definition of $\tau^{p+1} := \Pi^{\operatorname{div}} \tau$:

$$\begin{cases} \|\tau^{p+1} - \tau\| \to \min_{\tau^{p+1} \in V^{p+1}} \text{ subject to constraints (5.6.69)} \\ \text{and the constraint on the divergence: } \operatorname{div} \tau^{p+1} = y^p \,. \end{cases} \tag{5.6.72}$$

It follows from Lemma 5.19 that the problem is well-posed, provided we satisfy the minimum assumption on the enriched $H(\operatorname{div})$ test space,

$$\mathcal{RT}^{p+1}(T) \subset V^{p+1} \,,$$

and the divergence maps V^{p+1} *onto* space Y^p.

Theorem 5.21. *The operator defined by the constrained minimization problem* (5.6.72) *is well-defined, linear, and continuous,*

$$\Pi^{\operatorname{div}} : H(\operatorname{div}, K) \to V^{p+1}, \qquad \|\Pi^{\operatorname{div}} \tau\|_{H(\operatorname{div}, K)} \le C_{\Pi^{\operatorname{div}}} \|v\|_{H(\operatorname{div}, K)} \,.$$

The continuity constant $C_{\Pi^{\operatorname{div}}}$ is independent of element size, but it may depend upon the polynomial order p.

Proof. We begin by noting that the constraint on the divergence is compatible with but *not* independent of (5.6.69). According to Lemma 5.19, the first two constraints imply

$$(y^p - \operatorname{div} \tau, \chi)_K = 0 \quad \forall \chi \in \mathcal{P}^p(K) \,,$$

and, to enforce linear independence of the constraints, we can replace[24] the constraint on the divergence with the condition

$$(y^p - \operatorname{div} \tau, \chi)_K = 0 \quad \forall \chi \in \mathcal{P}^p(K)^\perp \,,$$

where $\mathcal{P}^p(K)^\perp$ is the L^2-orthogonal complement of $\mathcal{P}^p(K)$ in Y^p. The minimization problem can also be replaced with an equivalent orthogonality condition,

$$(\tau^{p+1} - \tau, \delta\tau)_K = 0 \quad \forall \delta\tau \in V \,,$$

where $V \subset V^{p+1}$ consists of all polynomials $\delta\tau$ satisfying the constraints

$$(\delta\tau, \psi)_K = 0 \quad \forall \psi \in \mathcal{P}^{p-1}(K)^3 \,,$$
$$\langle (\delta\tau) \cdot n.\phi \rangle_{\partial K} = 0 \quad \forall \phi \in \mathcal{P}^p(K) \,,$$
$$(\operatorname{div} \delta\tau, \chi)_K = 0 \quad \forall \chi \in Y^p \,.$$

It easily follows now that Fortin operator $\hat{\Pi}^{\operatorname{div}}$ is well-defined and linear with a null space closed in $H(\operatorname{div}, \hat{K})$. This implies continuity; compare Exercise 5.6.3. Finally, commuting property (5.6.68) implies the continuity of operator Π^{div} defined on an arbitrary affine tetrahedron K with a continuity constant independent of element size. $\square$

---

[24]Compare Exercise 5.6.6.

The constrained minimization problem (5.6.72) can be replaced with an equivalent mixed problem. Brezzi's theory then provides a more explicit control of the continuity constant for Π^{div} in terms of inf-sup constants for the forms defining the constraints; see Exercise 5.6.6.

We conclude this section by observing the action of operator Π^{div} on a curl, i.e., for $\tau = \mathrm{curl}\, F$. It follows from the construction that $\mathrm{div}(\Pi^{\mathrm{div}}\,\mathrm{curl}\, F) = 0$, so the constrained minimization problem to determine τ^{p+1} simplifies to

$$\|\tau^{p+1} - \mathrm{curl}\, F\| \to \min_{\tau^{p+1} \in V^{p+1}(\mathrm{div}_0)} \quad \text{subject to constraint (5.6.69)},\qquad (5.6.73)$$

where $V^{p+1}(\mathrm{div}_0)$ denotes the subspace of V^{p+1} of divergence-free functions.

5.6.3 ▪ Π^{curl} Fortin Operator

We follow the same logic as for the $H(\mathrm{div})$ operator, starting by defining the divergence of $\Pi^{\mathrm{curl}}F$. The obvious choice is to use operator (5.6.73), but we have to make a small correction accounting for the orthogonality property (5.6.56) involving polynomials of order p, one order higher than in (5.6.73). Thus, we seek $\tau^{p+2} := \mathrm{curl}\,\Pi^{\mathrm{curl}}F$ in the subspace of divergence-free functions from a larger space $V^{p+2} \supset \mathcal{RT}^{p+2}(K)$. In other words, we require that $\mathrm{curl}\, Q^{p+2} \supset \mathcal{P}^{p+1}(K)^3$. We have

$$\|\tau^{p+2} - \mathrm{curl}\, F\| \to \min_{\tau^{p+2} \in \mathrm{curl}\, Q^{p+2}} \quad \text{subject to constraints (5.6.56)}.\qquad (5.6.74)$$

We can now formulate a constrained minimization problem defining $\Pi^{\mathrm{curl}}F$,

$$\begin{cases} \Pi^{\mathrm{curl}} : H(\mathrm{curl}, K) \to Q^{p+2}, \quad \Pi^{\mathrm{curl}}F := F^{p+2} \in Q^{p+2}, \\[2mm] \|F^{p+2} - F\| \to \min_{F^{p+2} \in Q^{p+2}} \quad \text{subject to constraints (5.6.55) and the constraint on curl,} \\[2mm] \mathrm{curl}\, F^{p+2} = \tau^{p+2}. \end{cases}$$

$$(5.6.75)$$

It follows from Lemma 5.18 that the problem is well-posed, provided we satisfy the minimum assumption on the enriched $H(\mathrm{curl})$ test space:

$$\mathcal{N}^{p+2}(K) \subset Q^{p+2}.$$

Theorem 5.22. *The operator defined by the constrained minimization problem (5.6.75) is well-defined, linear, and continuous,*

$$\Pi^{\mathrm{curl}} : H(\mathrm{curl}, K) \to Q^{p+2}, \qquad \|\Pi^{\mathrm{curl}}F\|_{H(\mathrm{curl}, K)} \le C_{\Pi^{\mathrm{curl}}}\|F\|_{H(\mathrm{curl}, K)}.$$

The continuity constant $C_{\Pi^{\mathrm{curl}}}$ is independent of element size, but it may depend upon the polynomial order p.

Proof. The proof is analogous to the proof of Theorem 5.21. $\square$

We conclude this section by observing the action of operator Π^{curl} on a gradient, i.e., for $F = \nabla u$. It follows from the construction that $\mathrm{curl}(\Pi^{\mathrm{curl}}\nabla u) = 0$, so the constrained minimization problem to determine F^{p+2} simplifies to

$$\|F^{p+2} - \nabla u\| \to \min_{F^{p+2} \in Q^{p+2}(\mathrm{curl}_0)} \quad \text{subject to constraint } (5.6.55)_1,\qquad (5.6.76)$$

where $Q^{p+2}(\mathrm{curl}_0)$ denotes the subspace of Q^{p+2} of curl-free functions.

5.6.4 ▪ Π^{grad} Fortin Operator

By now, the reader should anticipate the construction and be able to fill in all necessary details. We seek $F^{p+3} := \nabla \Pi^{\mathrm{grad}} u$ in the subspace of curl-free functions from a larger space $Q^{p+3} \supset N^{p+3}(K)$. In other words, we require that $\nabla W^{p+3} \supset \mathcal{P}^{p+2}(K)^3$.

$$\|F^{p+3} - \nabla u\| \to \min_{F^{p+3} \in \nabla W^{p+3}} \quad \text{subject to constraints (5.6.65)}. \tag{5.6.77}$$

We now formulate a constrained minimization problem defining $\Pi^{\mathrm{grad}} u$,

$$\begin{cases} \Pi^{\mathrm{grad}} : H^1(K) \to W^{p+3}, \quad \Pi^{\mathrm{grad}} u := u^{p+3} \in W^{p+3}, \\ \|u^{p+3} - u\| \to \min_{u^{p+3} \in W^{p+3}} \text{subject to constraints: (5.6.64) and the constraint on gradient:} \\ \nabla u^{p+3} = F^{p+3}. \end{cases}$$

$$\tag{5.6.78}$$

It follows from Lemma 5.20 that the problem is well-posed, provided we satisfy the minimum assumption on the enriched H^1 test space:

$$\mathcal{P}^{p+3}(K) \subset W^{p+3}.$$

Theorem 5.23. *The operator defined by the constrained minimization problem (5.6.78) is well-defined and continuous,*

$$\Pi^{\mathrm{grad}} : H^1(K) \to W^{p+3}, \qquad \|\Pi^{\mathrm{grad}} u\|_{H^1(K)} \leq C_{\Pi^{\mathrm{grad}}} \|u\|_{H^1(K)}.$$

The continuity constant $C_{\Pi^{\mathrm{grad}}}$ is independent of element size, but it may depend upon the polynomial order p.

Exercises

5.6.1. Prove Lemma 5.20. *Hint:* Recall that if $\psi \in \mathcal{P}^{p-1}(K)$ with zero average, then there exists a polynomial $v \in \mathcal{P}^p(K)^3, v \cdot n = 0$ on ∂K, such that div $v = \psi$. (5 points)

5.6.2. Let $A : U \to V$ be a well-defined linear operator from a finite-dimensional normed space U into a normed space V. Show that A must be continuous. (2 points)

5.6.3. Let $A : U \to V$ be a well-defined linear operator from a normed space U into a normed space V, with a finite-dimensional range $\mathcal{R}(A) \subset V$. Show that A is continuous iff its null space $\mathcal{N}(A) \subset U$ is closed. (3 points)

5.6.4. Let $u = (u_1, u_2, u_3) \in U_1 \times U_2 \times U_3$ be a group variable where U_1, U_2, U_3 are Hilbert spaces. Consider a composite bilinear form,

$$b(u, v) := b_1(u_1, v) + b_2(u_2, v) + b_3(u_3, v),$$

where $v \in V$, a Hilbert test space. Define the kernel spaces

$$V_{12} := \{v \in V : b_1(u_1, v) + b_2(u_2, v) = 0 \quad u_1 \in U_1, u_2 \in U_2\},$$
$$V_1 := \{v \in V : b_1(u_1, v) = 0 \quad u_1 \in U_1\}$$

and assume three inf-sup conditions:

$$\sup_{v_{12} \in V_{12}} \frac{|b_3(u_3, v_{12})|}{\|v_{12}\|_V} \geq \gamma_3 \|u_3\|_{U_3},$$

$$\sup_{v_1 \in V_1} \frac{|b_2(u_2, v_1)|}{\|v_1\|_V} \geq \gamma_2 \|u_2\|_{U_2},$$

$$\sup_{v \in V} \frac{|b_1(u_1, v)|}{\|v\|_V} \geq \gamma_1 \|u_1\|_{U_1}.$$

Show that there exists a constant $\gamma = \gamma(\gamma_1, \gamma_2, \gamma_3, \|b_2\|, \|b_3\|)$ such that

$$\sup_{v \in V} \frac{|b(u, v)|}{\|v\|_V} \geq \gamma \left(\|u_1\|_{U_1}^2 + \|u_2\|_{U_2}^2 + \|u_3\|_{U_3}^2 \right)^{1/2}.$$

(3 points)

5.6.5. Prove Lemma 5.17. (3 points)

5.6.6. An alternate definition of Π^{div} operator. Prove that $\tau^{p+1} = \Pi^{\text{div}} \tau$ solves the following mixed problem:

$$\begin{cases} \tau^{p+1} \in V^{p+1}, \ \psi \in \mathcal{P}^{p-1}(K)^3, \ \phi \in \mathcal{P}_c^p(\partial K), \ \chi \in \mathcal{P}^p(K)^\perp, \\ (\tau^{p+1}, \delta\tau)_K + (\psi, \delta\tau)_K + \langle \phi, \delta\tau \rangle_{\partial K} + (\chi, \text{div}\,\delta\tau)_K \\ \qquad\qquad\qquad = (\tau, \delta\tau)_K, \qquad \delta\tau \in V^{p+1}, \\ (\delta\psi, \tau^{p+1})_K = (\delta\psi, \tau)_K, \qquad \delta\psi \in \mathcal{P}^{p-1}(K)^3, \\ \langle \delta\phi, \tau^{p+1} \cdot n \rangle_{\partial K} = \langle \delta\phi, \tau \cdot n \rangle_{\partial K}, \quad \delta\phi \in \mathcal{P}_c^p(\partial K), \\ (\delta\chi, \text{div}\,\tau^{p+1})_K = (\delta\chi, \text{div}\,\tau)_K, \quad \delta\chi \in \mathcal{P}^p(K)^\perp, \end{cases}$$
(5.6.79)

where $\mathcal{P}^p(K)^\perp$ is the L^2-orthogonal complement of $\mathcal{P}^p(K)$ in Y^p. Verify Brezzi conditions to prove that the mixed problem is well-posed and, therefore, the solution operator is well-defined and continuous. (10 points)

5.6.7. An alternate definition of Π^{curl} operator. Prove that constrained minimization problem (5.6.75) is equivalent to the mixed problem:

$$\begin{cases} F^{p+2} \in Q^{p+2}, \ \psi \in \mathcal{P}^{p-1}(K)^3, \ \phi \in \gamma_t(\mathcal{P}^p(K)^3), \ \tau \in V_0^{p+1}, \\ (F^{p+2}, \delta F)_K + (\psi, \delta F)_K + \langle n \times \phi, \delta F \rangle_{\partial K} + (\tau, \text{curl}\,\delta F)_K = (F, \delta F)_K, \\ \qquad\qquad\qquad\qquad\qquad\qquad\qquad\qquad\qquad\qquad \delta F \in Q^{p+2}, \\ (\delta\psi, F^{p+2})_K = (\delta\psi, F)_K, \qquad\qquad\qquad \delta\psi \in \mathcal{P}^{p-1}(K)^3, \\ \langle n \times \delta\phi, F^{p+2} \rangle_{\partial K} = \langle n \times \delta\phi, F \rangle_{\partial K}, \qquad \delta\phi \in \gamma_t(\mathcal{P}^p(K)^3), \\ (\delta\tau, \text{curl}\,F^{p+2})_K = (\delta\tau, \text{curl}\,F)_K, \qquad\qquad \delta\tau \in V_0^{p+1}, \end{cases}$$
(5.6.80)

where V_0^{p+1} is the L^2-orthogonal complement of the subspace of $\text{curl}\,Q^{p+2}$ satisfying constraints (5.6.56) in $\text{curl}\,Q^{p+2}$. Verify Brezzi's conditions to prove that the problem is well-posed and, therefore, operator Π^{curl} is well-defined and continuous. (10 points)

5.6.8. **An alternate definition of Π^{grad} operator.** Prove that constrained minimization problem (5.6.78) is equivalent to the mixed problem:

$$
\begin{cases}
u^{p+3} \in W^{p+3}, \ \psi \in \mathcal{P}^{p-1}(K)^3, \phi \in \gamma_n(\mathcal{P}^p(K)^3), \ \tau \in Q_0^{p+2}, \\[2mm]
(u^{p+3}, \delta u)_K + (\psi, \delta u)_K + \langle \phi, \delta u \rangle_{\partial K} + (F, \nabla \delta u)_K \\[2mm]
\qquad\qquad = (u, \delta u)_K, \qquad \delta u \in W^{p+3}, \\[2mm]
(\delta \psi, u^{p+3})_K = (\delta \psi, u)_K, \qquad \delta \psi \in \mathcal{P}^{p-1}(K)^3, \\[2mm]
\langle \delta \phi, u^{p+3} \rangle_{\partial K} = \langle \delta \phi, u \rangle_{\partial K}, \qquad \delta \phi \in \gamma_n(\mathcal{P}^p(K)^3), \\[2mm]
(\delta F, \nabla u^{p+3})_K = (\delta \tau, \nabla u)_K, \qquad \delta F \in Q_0^{p+2},
\end{cases}
$$

(5.6.81)

where Q_0^{p+2} is the L^2-orthogonal complement of the subspace of ∇W^{p+3} satisfying constraints (5.6.65) in ∇W^{p+3}. Verify Brezzi's conditions to prove that the problem is well-posed and, therefore, operator Π^{grad} is well-defined and continuous. (10 points)

5.7 ▪ The Double Adaptivity Method

In this section, we return to the Petrov–Galerkin method with optimal test functions in context of standard, conforming test spaces. The presentation follows closely [31]. As explained in the previous sections, the ideal scheme delivers the orthogonal projection in a special *energy norm*,

$$
\|u\|_E := \|Bu\|_{V'} = \sup_{v \in V} \frac{|b(u, v)|}{\|v\|_V}.
$$

Obviously, the energy norm depends upon the choice of test norm. Given now any suitable trial norm $\|u\|_U$ (consistent with the functional setting), it is natural to ask a question of whether we can find a test norm such that the corresponding energy norm will coincide with the trial norm. The answer is in principle "yes," and it is related to the concept of the so-called duality pairing.

Duality pairings. Let U, V be Hilbert spaces. A bilinear (sesquilinear) form $b(u, v)$, $u \in U, v \in V$, is called a *(generalized) duality pairing* if the following relations hold:

$$
\|u\|_U = \|b(u, \cdot)\|_{V'} = \sup_{v \in V} \frac{|b(u, v)|}{\|v\|_V} \quad \text{and} \quad \|v\|_V = \|b(\cdot, v)\|_{U'} = \sup_{u \in U} \frac{|b(u, v)|}{\|u\|_U}. \quad (5.7.82)
$$

In particular, the bilinear form is *definite*, i.e.,

$$
b(u, v) = 0 \quad \forall v \in V \quad \Rightarrow \quad u = 0 \qquad \text{and} \qquad b(u, v) = 0 \quad \forall u \in U \quad \Rightarrow \quad v = 0.
$$

This definition is motivated by the standard duality pairing, where $V = U'$, $b(u, v) = \langle u, v \rangle := v(u)$, and the (induced) norm in the dual space is defined by

$$
\|v\|_{U'} := \sup_{u \in U} \frac{|\langle u, v \rangle|}{\|u\|_U}.
$$

For nontrivial examples of duality pairings for trace spaces of the exact sequence energy spaces, see [27]. As in the case of the classical duality pairing, any definite bilinear (sesquilinear) form that satisfies the inf-sup condition *can be made in a duality pairing* if we equip V with the norm induced by the norm on U or, vice versa, space U with the norm induced by the norm on V.

That is, if we equip V with the norm induced by $\| \cdot \|_U$, then the induced norm on U equals the original norm on U,

$$\|v\|_V := \sup_{u \in U} \frac{|b(u,v)|}{\|u\|_U} \qquad \Rightarrow \qquad \sup_{v \in V} \frac{|b(u,v)|}{\|v\|_V} = \|u\|_U.$$

In the context of the Petrov–Galerkin method with optimal test functions, we call this test norm *the optimal test norm*,

$$\|v\|_{V_{\mathrm{opt}}} := \sup_{u \in U} \frac{|b(u,v)|}{\|u\|_U}. \tag{5.7.83}$$

To be of practical use, the optimal test norm must be computable. If we disregard 1D problems (see Exercise 5.7.1), the UW formulation stands out in the following way. Let

$$A : L^2(\Omega) \supset D(A) \to L^2(\Omega)$$

denote any well-defined closed operator corresponding to a system of first order PDEs with BCs included in the definition of its domain $D(A)$. Consider the boundary-value problem described by operator A,

$$\begin{cases} u \in D(A), \\ Au = f. \end{cases} \tag{5.7.84}$$

The UW formulation for the problem looks as follows:

$$\begin{cases} u \in L^2(\Omega), \\ \underbrace{(u, A^*v)}_{=:b(u,v)} = (f, v), \quad v \in D(A^*), \end{cases}$$

where A^* denotes the L^2-adjoint of operator A. If we choose the L^2-norm as the trial norm,[25] the optimal test norm for the UW formulation can be computed explicitly,

$$\|v\|_{V_{\mathrm{opt}}} = \sup_{u \in L^2(\Omega)} \frac{|(u, A^*v)|}{\|u\|} = \|A^*v\|.$$

We also refer to this norm as *the adjoint norm*. The corresponding adjoint graph norm, also called *the quasi-optimal test norm*, is defined by

$$\|v\|^2_{V_{\mathrm{qopt}}} := \|A^*v\|^2 + \alpha\|v\|^2, \quad \alpha > 0.$$

For the first order system (5.7.84) to be well-posed, operator A must be bounded below,

$$\|Au\| \geq \gamma\|u\| \qquad \Leftrightarrow \qquad \|A^*v\| \geq \gamma\|v\|.$$

The adjoint norm and the adjoint graph norm are then equivalent to each other,

$$\|A^*v\|^2 \leq \|A^*v\|^2 + \alpha\|v\|^2 \qquad \text{and} \qquad \|A^*v\|^2 + \alpha\|v\|^2 \leq \left(1 + \frac{\alpha}{\gamma^2}\right)\|A^*v\|^2. \tag{5.7.85}$$

The corresponding energy norms are then equivalent to each other as well since

$$C_1\|v\|_{V_2} \leq \|v\|_{V_1} \leq C_2\|v\|_{V_2}$$

---

[25] We also could choose a *weighted* L^2-norm.

implies

$$\frac{1}{C_2} \sup_v \frac{|b(u,v)|}{\|v\|_{V_2}} \leq \sup_v \frac{|b(u,v)|}{\|v\|_{V_1}} \leq \frac{1}{C_1} \sup_v \frac{|b(u,v)|}{\|v\|_{V_2}}.$$

As we try to keep the equivalence constant in (5.7.85) close to one, this suggests selecting scaling constant α in the adjoint graph norm to be of order γ^2. The name *quasi-optimal test norm* is justified by the fact that the method with the adjoint graph norm delivers an orthogonal projection in a norm equivalent to the L^2 trial norm.

Three mixed problems.   Consider our usual abstract variational problem,

$$\begin{cases} u \in U, \\ b(u,v) = l(v), \quad v \in V. \end{cases}$$

Instead of discretizing the problem directly, we follow the approach proposed by Cohen, Dahmen, and Welper [20], and replace it with a mixed problem,

$$\begin{cases} \psi \in V, u \in U, \\ (\psi,v)_V + b(u,v) = l(v), \quad v \in V, \\ \quad\quad b(\delta u, \psi) = 0, \quad\quad \delta u \in U. \end{cases} \tag{5.7.86}$$

Function $\psi \in V$ is identified as the *Riesz representation of the residual*,

$$(\psi,v)_V = l(v) - b(u,v), \quad v \in V,$$

and, on the continuous level, is zero. Obviously, both formulations deliver the same solution u. This is no longer true on the approximate level. The *ideal Petrov–Galerkin (PG) method with optimal test functions* seeks an approximate solution $\tilde{u}_h \in U_h$ along with the corresponding exact (Riesz representation of) residual $\psi^h \in V$ that solves the semidiscrete mixed problem:

$$\begin{cases} \psi^h \in V, \tilde{u}_h \in U_h, \\ (\psi^h,v)_V + b(\tilde{u}_h,v) = l(v), \quad v \in V, \\ \quad\quad b(\delta u_h, \psi^h) = 0, \quad\quad \delta u_h \in U_h. \end{cases} \tag{5.7.87}$$

The ideal PG method with optimal test functions delivers orthogonal projection $\tilde{u}_h$ in the energy norm.

For obvious reasons, we cannot compute with the ideal PG method. We need to approximate space V with some finite-dimensional subspace $V_h \subset V$. The ultimate approximate problem reads then as follows:

$$\begin{cases} \psi_h \in V_h, u_h \in U_h, \\ (\psi_h,v_h)_V + b(u_h,v_h) = l(v_h), \quad v_h \in V_h, \\ \quad\quad b(\delta u_h, \psi_h) = 0, \quad\quad \delta u_h \in U_h. \end{cases} \tag{5.7.88}$$

This is the *practical PG method with optimal test functions*. Brezzi's theory tells us that we have to now satisfy two discrete inf-sup conditions. The *inf-sup in kernel* is trivially satisfied because of the presence of the test inner product. The discrete LBB condition,

$$\sup_{v_h \in V_h} \frac{|b(u_h,v_h)|}{\|v_h\|_V} \geq \gamma \|u_h\|_U, \quad u_h \in U_h,$$

coincides with the discrete Babuška condition for the original problem but *it is much easier now to satisfy it as we can employ test spaces of larger dimension*:

$$\dim V_h \gg \dim U_h \,.$$

The *discontinuous* PG method employs variational formulations with *discontinuous* (broken, product) test spaces, and the standard way to guarantee the satisfaction of the discrete LBB condition has been to use *enriched* test spaces with order $r = p + \Delta p$ where p is the polynomial order of the trial space, and $\Delta p > 0$ is an increment in the order of approximation.

Double adaptivity.   The groundbreaking idea of Cohen, Dahmen, and Welper [20] is to determine an optimal discrete test space V_h using adaptivity. After all, the fully discrete mixed problem (5.7.88) is supposed to be an approximation of the semidiscrete mixed problem (5.7.87). Both problems share the same discrete trial space U_h and the task is to determine a good approximation $\psi_h \in V_h$ to the ideal $\psi^h \in V$ in terms of the test norm. This, hopefully, should guarantee that the corresponding ultimate discrete solution $u_h \in U_h$ approximates the ideal discrete solution $\tilde{u}_h \in U_h$ as well. We arrive at the concept of the double adaptivity algorithm described below.

Given error tolerances $\mathrm{tol}_U, \mathrm{tol}_V$ for the trial and test mesh, we proceed as follows:

ALGORITHM 5.1.
Double adaptivity

```
Set initial trial mesh Uₕ
do
   (re)set the test mesh Vₕ to coincide with the trial mesh Uₕ
   do
      solve problem (5.7.88) on the current meshes
      estimate error err_V := ‖ψʰ − ψₕ‖_V and compute norm ‖ψₕ‖_V
      if err_V/‖ψₕ‖_V < tol_V exit the inner (test) loop
      adapt the test mesh Vₕ using element contributions of err_V
   enddo
   compute trial norm of the solution ‖uₕ‖_U
   if ‖ψₕ‖_V/‖uₕ‖_U < tol_U STOP
   use element contributions to ‖ψₕ‖_V to refine the trial mesh
enddo
```

A few comments are in place.   By setting the test mesh to the trial mesh, we mean the corresponding mesh data structure. The trial and test energy spaces may be different, dependent upon the variational formulation. There are two main challenges in implementing this method. The first one is on the coding side. As the logic of double adaptivity calls for two independent meshes, developing an adaptive code in this context seems to be very nontrivial. In our *hp*-adaptive finite element code, written in Fortran, we have resolved this problem by using *pointers* to separate mesh data structures. This way, the adaptivity code, conceptualized for one mesh, can be extended to support two or more *independent meshes*. The second challenge lies in developing a reliable a posteriori error estimation technique for the inner (test) adaptivity loop. After several unsuccessful attempts we have converged to a duality technique described here. It is in the context of the duality-based error estimation that the UW variational formulation distinguishes itself from other formulations one more time.

Duality theory. We now discuss the main technical issue in this section: the a posteriori error estimation and adaptivity for the inner loop problem based on the classical duality theory [41].

We begin by noticing that the semidiscrete mixed problem (5.7.87) is equivalent to the constrained minimization (primal) problem:

$$\inf_{\substack{\psi \in D(A^*) \\ A^*\psi \in U_h^\perp}} \underbrace{\frac{1}{2}\|A^*\psi\|^2 + \frac{\alpha}{2}\|\psi\|^2 - (f, \psi)}_{=:J(\psi)}, \tag{5.7.89}$$

where $U_h^\perp$ denotes the $L^2(\Omega)$-orthogonal complement of discrete trial space U_h. Next, we introduce an auxiliary variable,

$$\sigma = A^*\psi.$$

Recalling density of $D(A)$ in $L^2(\Omega)$[26], we have

$$\sup_{v \in L^2(\Omega)} (A^*u - \sigma, v) = \sup_{v \in D(A)} (A^*u - \sigma, v) = \begin{cases} 0 & \text{if } A^*u = \sigma, \\ +\infty & \text{otherwise.} \end{cases}$$

Consequently, we can turn the minimization problem into a saddle point problem,

$$\begin{aligned}
&\inf_{\substack{\psi \in D(A^*) \\ A^*\psi \in U_h^\perp}} \underbrace{\frac{1}{2}\|A^*\psi\|^2 + \alpha\frac{1}{2}\|\psi\|^2 - (f, \psi)}_{=:J(\psi)} \\
&= \inf_{\substack{\sigma \in L^2(\Omega) \\ \sigma \in U_h^\perp}} \inf_{\psi \in D(A^*)} \sup_{\phi \in D(A)} \left\{ \frac{1}{2}\|\sigma\|^2 + \alpha\frac{1}{2}\|\psi\|^2 - (f, \psi) + (A^*\psi - \sigma, \phi) \right\} \\
&= \inf_{\substack{\sigma \in L^2(\Omega) \\ \sigma \in U_h^\perp}} \inf_{\psi \in D(A^*)} \sup_{\phi \in D(A)} \left\{ \frac{1}{2}\|\sigma\|^2 + \alpha\frac{1}{2}\|\psi\|^2 - (f, \psi) + (\psi, A\phi) - (\sigma, \phi) \right\} =: (^*).
\end{aligned}$$

$$(5.7.90)$$

At this point, we are ready to trade the inf-sup for the sup-inf,

$$(^*) \geq \sup_{\phi \in D(A)} \inf_{\substack{\sigma \in L^2(\Omega) \\ \sigma \in U_h^\perp}} \inf_{\psi \in D(A^*)} \left\{ \frac{1}{2}\|\sigma\|^2 + \alpha\frac{1}{2}\|\psi\|^2 - (f, \psi) + (\psi, A\phi) - (\sigma, \phi) \right\} =: (^{**}).$$

We plan to show a posteriori that, in fact, we still have the equality above.

The whole point is that we can now compute the two minimization problems *explicitly*. Minimization in σ yields

$$\sigma = \phi^\perp \quad \Rightarrow \quad \inf_{\substack{\sigma \in L^2(\Omega) \\ \sigma \in U_h^\perp}} \frac{1}{2}\|\sigma\|^2 - (\sigma, \phi) = -\frac{1}{2}\|\phi^\perp\|^2.$$

Minimizing in $\psi \in D(A^*)$, we get

$$\alpha\psi = f - A\phi \quad \Rightarrow \quad \inf_{\psi \in D(A^*)} \left\{ \frac{\alpha}{2}\|\psi\|^2 - (f - A\phi, \psi)+ \right\} = -\frac{1}{2\alpha}\|f - A\phi\|^2.$$

In the end, we obtain the dual problem

$$(^{**}) = \sup_{\phi \in D(A)} \underbrace{-\frac{1}{2}\|\phi^\perp\|^2 - \frac{1}{2\alpha}\|f - A\phi\|^2}_{=:J^*(\phi)}. \tag{5.7.91}$$

---

[26] A necessary assumption for introducing the adjoint; see [66].

Simple algebra and integration by parts show that

$$2(J(\psi) - J^*(\phi)) = \frac{1}{\alpha} \int_\Omega \{\alpha(A^*\psi - \phi^\perp)^2 + (\alpha\psi - (f - A\phi))^2\}$$

for *any* $\psi \in D(A^*)$ and $\phi \in D(A)$. Next, we demonstrate that if ψ is the solution of the primal minimization problem and ϕ is the solution of the dual maximization problem, then the right-hand side above is equal to zero, i.e., there is no duality gap on the continuous level. Naturally, this is necessary to later use the duality gap for the a posteriori error estimation for approximate solutions. Strict convexity of the primal functional and strict concavity of the dual functional imply that the minimizers of $J(\psi)$ and $-J^*(\phi)$ exist and are unique.

The solution of the primal problem satisfies the mixed problem

$$\begin{cases} \psi \in D(A^*),\ \tilde{u}_h \in U_h\,, \\ (A^*\psi, A^*\delta\psi) + (\alpha\psi, \delta\psi) + (\tilde{u}_h, A^*\delta\psi) = (f, \delta\psi), & \delta\psi \in D(A^*)\,, \\ (A^*\psi, \delta\tilde{u}_h) = 0, & \delta\tilde{u}_h \in U_h\,, \end{cases} \qquad (5.7.92)$$

where $\tilde{u}_h \in U_h$ is the corresponding Lagrange multiplier.

The solution to the dual problem satisfies another mixed problem:

$$\begin{cases} \phi \in D(A),\ \tilde{w}_h \in U_h\,, \\ (A\phi, A\delta\phi) + \alpha(\phi, \delta\phi) - \alpha(\tilde{w}_h, \delta\phi) = (f, A\delta\phi), & \delta\phi \in D(A)\,, \\ -\alpha(\phi, \delta w_h) + \alpha(\tilde{w}_h, \delta w_h) = 0, & \delta w_h \in U_h\,, \end{cases}$$

or, in the strong form,

$$A^*A\phi + \alpha(\phi - \tilde{w}_h) = A^*f \qquad (5.7.93)$$

plus the BC,

$$BA\phi = Bf \quad \Rightarrow \quad f - A\phi \in D(A^*)\,, \qquad (5.7.94)$$

where boundary operator B corresponds to BCs built into the definition of $D(A)$. Now let ϕ be the solution to the dual problem. Use one of the duality relations to define *a function* ψ,

$$\psi := \frac{1}{\alpha}(f - A\phi)\,.$$

First of all, ψ satisfies the second duality relation. Indeed, (5.7.93) implies that

$$A^*\psi = \frac{1}{\alpha}(A^*f - A^*A\phi) = \phi - \tilde{w}_h = \phi^\perp\,.$$

Second, BC (5.7.94) implies that $\psi \in D(A^*)$. Finally, plugging the function ψ and $\tilde{u}_h = \tilde{w}_h$ into variational formulation (5.7.92)$_1$, we obtain

$$\begin{aligned} (A^*\psi, A^*\delta u) + (\alpha\psi, \delta u) + (\tilde{u}_h, A^*\delta u) &= (\phi^\perp, A^*\delta u) + (f - A\phi, \delta u) + (\tilde{w}_h, A^*\delta u) \\ &= (\phi, A^*\delta u) - (A\phi, \delta u) + (f, \delta u) \\ &= (f, \delta u)\,. \end{aligned}$$

Note that the duality relation $A^*\psi = \phi^\perp$ implies that (5.7.92)$_2$ is satisfied as well. Consequently, uniqueness of the solution to the primal problem implies that function ψ derived from the duality relations indeed is the solution of the primal problem. *There is no duality gap on the continuous level.*

A posteriori error estimation.   Solving the primal and dual problems approximately for ψ_h and ϕ_h, we can use the duality gap $2(J(\psi_h) - J^*(\phi_h))$ to estimate the error in the energy norms,

$$\left.\begin{array}{c} \frac{1}{\alpha}\left\{\alpha\|A^*(\psi - \psi_h)\|^2 + \|\psi - \psi_h\|^2\right\} \\ \frac{1}{\alpha}\left\{\alpha\|\phi^\perp - \phi_h^\perp\|^2 + \|A(\phi - \phi_h)\|^2\right\} \end{array}\right\} \leq 2(J(\psi_h) - J^*(\phi_h)),$$

where the duality gap can be expressed in terms of the integral of the consistency terms,

$$2(J(\psi_h) - J^*(\phi_h)) = \frac{1}{\alpha}\int_\Omega \alpha(A^*\psi_h - \sigma_h)^2 + (\alpha\psi_h - (f - A\phi_h))^2. \tag{5.7.95}$$

Element contributions,

$$\int_K \alpha(A^*\psi_h - \sigma_h)^2 + (\alpha\psi_h - (f - A\phi_h))^2,$$

will serve as element error indicators.

Remark 5.24. *Can we pass with $\alpha \to 0$?* Clearly, for small α, the dual problem approaches the least squares method for the original problem, and the least squares term dominates the duality gap. The two problems disconnect, and the duality gap is no longer a meaningful estimate for either the primal or the dual problem. This is consistent with the well-known fact that the duality theory for linear elastostatics requires the maximization over stress fields satisfying the equilibrium equations. In our case, we would need to maximize over ϕ_h satisfying the equation $A\phi = f$. There is only one such ϕ—the solution to our problem. In conclusion, *we have to compute with positive α.*

Controlling the error.   The ideal PG method (with infinite-dimensional test space) inherits the inf-sup condition from the continuous problem. In other words, the operator $B : U \to V'$ generated by the bilinear form $b(u, v)$ is bounded below. This implies that the error $u - \tilde{u}_h$ is controlled by the residual,

$$\gamma\|u - \tilde{u}_h\|_U \leq \|l - B\tilde{u}_h\|_{V'} = \|\psi^h\|_V.$$

Once the residual converges to zero, so must the error, at the same rate. The inner adaptivity loop guarantees that we approximate the (Riesz representation of) residual ψ^h within a required tolerance with ψ_h. But with ψ_h, only the approximation u_h of $\tilde{u}_h$ is available. How do we know that u_h converges to $\tilde{u}_h$? Can we estimate the difference $\tilde{u}_h - u_h$? This question deals again with a mixed problem, albeit one where space U_h is finite-dimensional. An attempt to use Brezzi's theory makes little sense as it calls for a discrete LBB inf-sup condition, which is precisely what we are trying to circumvent.

This is where the duality theory comes into play again. The critical piece of information is that the ideal approximate solution $\tilde{u}_h$ coincides with the L^2-projection $\tilde{w}_h$ of the solution ϕ of the dual problem (see the reasoning above showing that there is no duality gap for the exact ϕ and ψ). The primal problem is a standard[27] mixed problem but the dual problem is a (double) minimization problem. The duality gap used to estimate the error in the approximate solution to the primal problem also estimates the error in the solution of the dual problem,

$$\|A(\phi - \phi_h)\|^2 + \alpha\|\phi^\perp - \phi_h^\perp\|^2 \leq 2(J(\psi_h) - J^*(\phi_h)) =: \text{est}.$$

---

[27]Originating from a constrained minimization problem.

Operator A is bounded below with a constant β,

$$\beta\|\phi - \phi_h\| \le \|A(\phi - \phi_h)\|,$$

which implies that

$$\beta^2\|\phi - \phi_h\|^2 \le \text{est}.$$

This in turn implies the bound for the projection as well,

$$\beta^2\|\tilde{w}_h - w_h\|^2 \le \beta^2\|\phi - \phi_h\|^2 \le \text{est}. \tag{5.7.96}$$

In summary, we should use w_h and not u_h as our final (numerical) solution of the problem.

5.7.1 ▪ Example: Confusion Problem

In the entire book so far, we have avoided numerical illustrations of the presented theories, referring to related papers and other books. In this section, we make an exception to this rule and illustrate the double adaptivity algorithm with a simple 1D numerical example. Please pay attention to the numerical results to develop a good understanding of the discussed concepts.

Consider the convection-dominated diffusion problem, in short, the *confusion problem*:

$$\begin{cases} u = 0 & \text{on } \Gamma, \\ -\epsilon\Delta u + \beta \cdot \nabla u = f & \text{in } \Omega, \end{cases} \tag{5.7.97}$$

where ϵ is a diffusion constant, and β denotes an advection vector. We begin by rewriting the second order problem as a system of first order equations. This can be done in more than one way. We will use the formulation advocated by Broersen and Stevenson [12, 13]:

$$\begin{cases} u = 0 & \text{on } \Gamma, \\ \sigma - \epsilon^{\frac{1}{2}}\nabla u = 0 & \text{in } \Omega, \\ -\epsilon^{\frac{1}{2}}\text{div}\,\sigma + \beta \cdot \nabla u = f & \text{in } \Omega. \end{cases}$$

The first equation defines the auxiliary variable—a scaled viscous flux. Splitting the diffusion constant ϵ in between the two equations is motivated by a better control of the round-off error for very small ϵ. More importantly, one can show that the corresponding first order operator is *robustly bounded below*, i.e., the boundedness below constant is *independent of* ϵ; see [31].

We now introduce the formalism of closed operators theory. Introducing the first order operator and its L^2-adjoint,

$$\begin{aligned} &\mathsf{u} := (\sigma, u) \in D(A) := H(\text{div}, \Omega) \times H_0^1(\Omega) \subset (L^2(\Omega))^N \times L^2(\Omega) \stackrel{'}{=} L^2(\Omega), \\ &A : D(A) \to L^2(\Omega), Au = A(\sigma, u) := (\sigma - \epsilon^{\frac{1}{2}}\nabla u, -\epsilon^{\frac{1}{2}}\text{div}\,\sigma + \beta \cdot \nabla u), \\ &\mathsf{v} := (\tau, v) \in D(A^*) = D(A), \\ &A^* : D(A^*) \to L^2(\Omega), A^*\mathsf{v} = A^*(\tau, v) = (\tau + \epsilon^{\frac{1}{2}}\nabla v, \epsilon^{\frac{1}{2}}\text{div}\,\tau - \text{div}(\beta v)), \end{aligned} \tag{5.7.98}$$

we can rewrite the problem in a concise form as

$$\begin{cases} \mathsf{u} \in D(A), \\ A\mathsf{u} = \mathsf{f}, \end{cases}$$

where $\mathsf{f} = (0, f)$.

Multiplying the equation with a test function $v = (\tau, v) \in D(A^*)$ and integrating by parts, we obtain the UW formulation:

$$\begin{cases} u \in L^2(\Omega), \\ (u, A^*v) = (f, v), \quad v \in D(A^*). \end{cases} \quad (5.7.99)$$

Numerical experiments. We now present 1D numerical experiments for $\Omega = (0,1), \beta = 1, f = 1$.

We start with a moderate value of $\epsilon = 10^{-2}$ to illustrate the algorithm. Our original trial mesh consists of five cubic elements, and the starting test mesh is set to the trial mesh but with elements of one order higher. Note that by the order of elements we mean always the order for the H^1-conforming elements in the 1D exact sequence. This means that effectively we approximate σ and u with piecewise quadratics, and the two components of residual ψ with piecewise quartic elements. Raising the initial order of test functions is related to the use of a classical frontal solver without pivoting. For $p = 1$, and trial and test meshes of equal order, we encounter a zero pivot in the very first element. The tolerance for the outer and inner loop adaptivity is set to 1% and 5% percent, respectively. We use the Dörfler refinement strategy [40] with 1% and 25% factors. The first inner loop iterations (a total of 9) are presented in Figs. 5.3 and 5.4. The solutions seem to evolve very little but the a posteriori error estimate evolves from a 162% to 4.7% relative error; see Table 5.1. Note that the ultimate discrete solution is *not* the L^2-projection of the exact solution. This is a consequence of using the adjoint graph norm rather than the adjoint norm.

The evolution of "trusted" trial solutions along with the corresponding resolved residual is shown in Fig. 5.5. In order to solve the problem with the requested 1 percent of accuracy, the algorithm has performed five outer loop iterations. The corresponding evolution of the relative error and inner loop duality error estimates is shown in Table 5.1.

1	50.6	162.4	76.8	35.9	23.6	14.4	9.0	5.6	4.7
2	27.8	106.3	34.3	20.1	10.3	7.2	5.0	2.7	
3	10.9	59.7	12.1	8.5	4.2				
4	2.6	31.0	4.6						
5	0.4	21.4	16.4	9.7	4.3				

Table 5.1: UW formulation, $\epsilon = 10^{-2}$. Column 1: outer loop iteration number. Column 2: error (residual) estimate for the "trusted" solution. Column 3 and next: evolution of inner loop a posteriori error estimate.

A couple of simple observations: (a) The number of inner loop iterations decreases with each outer loop iteration. (b) The residual for the unresolved solution has a significant variation, not only in the boundary layer but also at the inflow. At the end, the residual around the inflow becomes insignificant—note the lack of refinements at the inflow in the last test mesh. Conceptually, we need to think of a new residual after each trial mesh refinement. If we decide to keep the test mesh from the previous inner loop iterations, we need to implement unrefinements.

Convergence of u_h. To illustrate the point about the convergence of u_h to $\tilde{u}_h$, we present approximate solution u_h and projection w_h (second components) at the beginning and at the end of the first inner loop; see Fig. 5.6. As we can see, with an unresolved residual, the two functions are significantly different. However, once the residual has been resolved (error tolerance = 5%), the two solutions are indistinguishable.

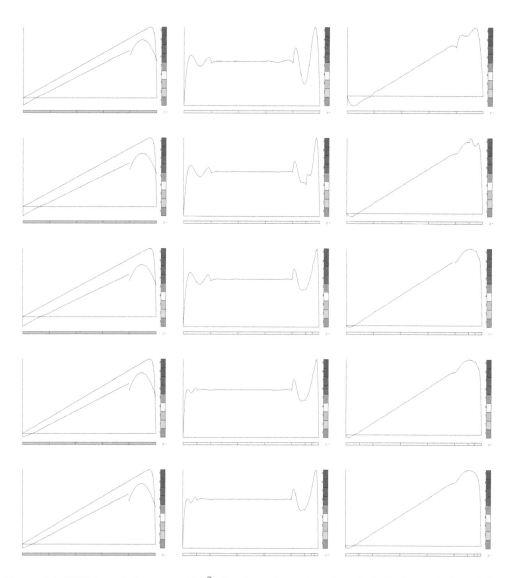

Figure 5.3: UW formulation, $\epsilon = 10^{-2}$, first inner loop, iterations 1–5. Left: evolution of the approximate solution u_h on a trial mesh of five cubic elements corresponding to different test meshes. Middle: the test mesh with the corresponding u component of approximate residual ψ_h. Right: the test mesh with the corresponding v component of the solution to the dual problem.

Performance of the method for small diffusion. We have been able to solve the problem for $\epsilon = 10^{-6}$ but we failed for $\epsilon = 10^{-7}$. The number of inner loop iterations increased significantly with smaller ϵ, and in the end, the inner loop iterations did not converge. We have implemented a verification based on a number of energy identities which should be satisfied (with infinite precision), and the code stopped passing those tests. Note that the duality gap estimate *has to decrease* with any mesh refinements. This stopped being the case in the end of the last run. In this case, the numerical errors grow too large because calculating with the square of the diffusivity constant ϵ cannot be performed with sufficient accuracy using double precision arithmetic.

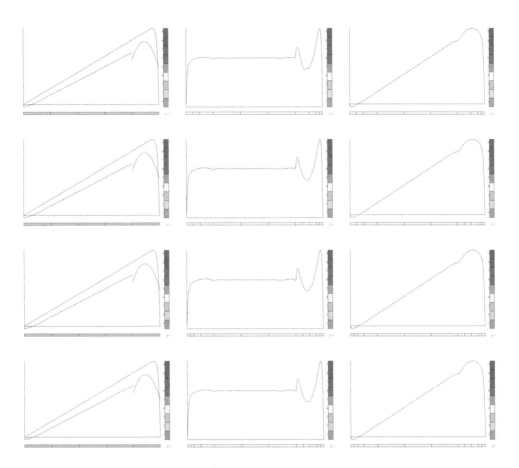

Figure 5.4: UW formulation, $\epsilon = 10^{-2}$, first inner loop, iterations 6–9. Left: evolution of the approximate solution u_h on a trial mesh of five cubic elements corresponding to different test meshes. Middle: the test mesh with the corresponding u component of approximate residual ψ_h. Right: the test mesh with the corresponding v component of solution to the dual problem.

$\epsilon = 10^{-6}$	33	32	32	31	31	31	30	33	47	45	42	39	37	37	38	41	30	14		
$\epsilon = 10^{-7}$	42	41	41	40	40	40	40	40	40	52	57	56	53	50	47	46	45	46	48	*

Table 5.2: UW formulation. Number of inner loop iterations for very small values of the diffusion constant. The star indicates no convergence.

We had more success using continuation in ϵ. Starting with $\epsilon = 10^{-2}$, we run the double adaptivity algorithm. Upon convergence, we restart the algorithm with $\epsilon_{\text{new}} = \epsilon_{\text{old}}/2$ and the initial trial mesh obtained from the previous run. Except for the last couple of cases, the number of inner loop iterations dramatically decreases (it did not exceed 10) and, ultimately, the smallest value of ϵ for which we have been able to solve the problem was $\epsilon = 3.814 10^{-8}$.

LBB condition and robustness. For each trial mesh U_h, the inner adaptivity algorithm produces the corresponding discrete *trusted test mesh* V_h that guarantees the resolution of the residual with a prescribed tolerance (5% in our numerical experiment). The residual and the test mesh

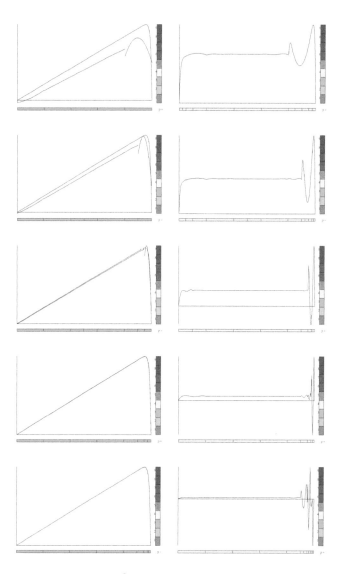

Figure 5.5: UW formulation, $\epsilon = 10^{-2}$, outer loop, iterations 1–5. Left: evolution of the approximate solution u_h. Right: the test mesh with the corresponding resolved u component of approximate residual ψ_h.

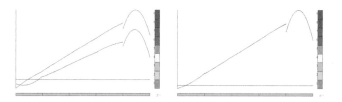

Figure 5.6: UW formulation, $\epsilon = 10^{-2}$, first inner loop. Second components of projection w_h (top) and approximate solution u_h (bottom). Left: at the beginning of the inner loop. Right: at the end of the loop.

correspond to a *particular load* (function f) and there is no reason to believe that the discrete inf-sup constant corresponding to the two meshes is ϵ-independent. This would have guaranteed robust stability *for an arbitrary load.* The method delivers a solution to a singular perturbation problem *without guaranteeing the robust discrete stability.*

Remark 5.25. If boundedness below constant β depends upon ϵ, then, unfortunately, bound (5.7.96) is *not* robust in ϵ, even if we choose α to be of order β^2. This is *not the case* for the Broersen and Stevenson formulation, for which it has been shown that β *is independent of* ϵ; see [31].

Exercises

5.7.1. Computation of the optimal test norm for 1D problems. Consider the classical variational formulation for the 1D confusion problem,

$$\begin{cases} u \in H_0^1(0,1)\,, \\ \epsilon(u',v') + (u',v) = (f,v)\,, \quad v \in H_0^1(0,1)\,. \end{cases}$$

Equip the trial space with the H_0^1-norm,

$$\|u\|_U := \|u'\|\,,$$

and show that the corresponding optimal test norm is as follows:

$$\|v\|_{V_{\mathrm{opt}}}^2 = \epsilon^2\|v'\|^2 + \|v\|^2 - \left(\int_0^1 v\right)^2\,.$$

Can you build a DPG method with such a test norm? (5 points)

5.7.2. Repeat Exercise 5.7.1 for the same 1D confusion equation but with a flux BC at $x = 0$,

$$\begin{cases} -\epsilon u'' + u' = f & \text{in } (0,1)\,, \\ -\epsilon u' + u = 0 & \text{at } x = 0\,, \\ u = 0 & \text{at } x = 1\,. \end{cases}$$

Derive the corresponding classical variational formulation; assume the H_0^1 trial norm and show that the corresponding optimal test norm is given by

$$\|v\|_{V_{\mathrm{opt}}}^2 = \|\epsilon v' - v + v(0)\|^2\,.$$

Can you build a DPG method with *this* test norm? (5 points)

Bibliography

[1] D. N. Arnold. Spaces of finite element differential forms. Analysis and numerics of partial differential equations. In *Analysis and numerics of partial differential equations*, INdAM Ser., pages 117–140. Springer, Milan, 2013. (Cited on p. 53)

[2] D. N. Arnold. *Finite Element Exterior Calculus*. SIAM, 2018. (Cited on p. 53)

[3] D. N. Arnold, R. S. Falk, and R. Winther. Preconditioning in $H(\mathrm{div})$ and applications. *Math. Comp.*, 66(219):957–984, 1997. (Cited on pp. 102, 103)

[4] I. Babuška. Error bounds for finite element method. *Numer. Math.*, 16:322–333, 1971. (Cited on p. 115)

[5] I. Babuška, R. B. Kellogg, and J. Pitkäranta. Direct and inverse error estimates for finite elements with mesh refinements. *Numer. Math.*, 33:447–471, 1979. (Cited on pp. 137, 138, 144)

[6] J. W. Barret and K. W. Morton. Approximate symmetrization and Petrov-Galerkin methods for diffusion-convection problems. *Comput. Methods Appl. Mech. Engrg.*, 46:97–122, 1984. (Cited on p. 148)

[7] D. Boffi, F. Brezzi, and M. Fortin. *Mixed Finite Element Methods and Applications*, volume 44 of *Springer Series in Computational Mathematics*. Springer, 2013. (Cited on p. 126)

[8] D. Boffi, M. Costabel, M. Dauge, L. Demkowicz, and R. Hiptmair. Discrete compactness for the p-version of discrete differential forms. *SIAM J. Numer. Anal.*, 49(1):135–158, 2011. (Cited on p. 135)

[9] C. L. Bottasso, S. Micheletti, and R. Sacco. The discontinuous Petrov-Galerkin method for elliptic problems. *Comput. Methods Appl. Mech. Engrg.*, 191:3391–3409, 2002. (Cited on p. 147)

[10] S. C. Brenner and L. R. Scott. *The Mathematical Theory of Finite Element Methods*. Springer-Verlag, 2010. Third Edition. (Cited on pp. xi, 39)

[11] F. Brezzi. On the existence, uniqueness and approximation of saddle-point problems arising from Lagrange multipliers. *RAIRO*, 8(R2):129–151, 1974. (Cited on pp. 126, 129)

[12] D. Broersen and R. P. Stevenson. A robust Petrov-Galerkin discretisation of convection-diffusion equations. *Comput. Math. Appl.*, 68(11):1605–1618, 2014. (Cited on p. 190)

[13] D. Broersen and R. P. Stevenson. A Petrov-Galerkin discretization with optimal test space of a mild-weak formulation of convection-diffusion equations in mixed form. *IMA J. Numer. Anal.*, 35(1):39–73, 2015. (Cited on p. 190)

[14] W. Cao and L. Demkowicz. Optimal error estimate for the projection-based interpolation in three dimensions. *Comput. Math. Appl.*, 50:359–366, 2005. (Cited on pp. 60, 82)

[15] C. Carstensen. Clément interpolation and its role in adaptive finite element error control. In *Partial Differential Equations and Functional Analysis*, volume 168 of *Operator Theory: Advances and Applications*, pages 27–43. Springer, 2006. (Cited on p. 107)

197

[16] C. Carstensen, L. Demkowicz, and J. Gopalakrishnan. Breaking spaces and forms for the DPG method and applications including Maxwell equations. *Comput. Math. Appl.*, 72(3):494–522, 2016. (Cited on pp. 154, 172)

[17] P. Causin and R. Sacco. A discontinuous Petrov–Galerkin method with Lagrangian multipliers for second order elliptic problems. *SIAM J. Numer. Anal.*, 43(1):280–302, 2005. (Cited on p. 147)

[18] Ph. G. Ciarlet. *The Finite Element Methods for Elliptic Problems*. North-Holland, 1994. (Cited on pp. xi, 53, 55)

[19] Ph. Clément. Approximation by finite element functions using local regularization. *RAIRO Anal. Numér.*, 9:77–84, 1975. (Cited on p. 107)

[20] A. Cohen, W. Dahmen, and G. Welper. Adaptivity and variational stabilization for convection-diffusion equations. *ESAIM Math. Model. Numer. Anal.*, 46(5):1247–1273, 2012. (Cited on pp. 147, 185, 186)

[21] M. Costabel and M. Dauge. On the inequalities of Babuška–Aziz, Friedrichs and Horgan–Payne. *Arch. Ration. Mech. Anal.*, 217:873–898, 2015. (Cited on p. 131)

[22] J. A. Cottrell, T. J. R. Hughes, and Y. Bazilevs. *Isogeometric Analysis: Toward Integration of CAD and FEA*. Wiley, 2009. (Cited on p. 59)

[23] L. Demkowicz. Asymptotic convergence in finite and boundary element methods. Part 1: Theoretical results. *Comput. Math. Appl.*, 27(12):69–84, 1994. (Cited on pp. 120, 123)

[24] L. Demkowicz. Projection-based interpolation. In *Transactions on Structural Mechanics and Materials*. Monograph 302. Cracow University of Technology Publications, Cracow, 2004. See also *ICES Report* 04-03. (Cited on pp. 60, 82)

[25] L. Demkowicz. *Computing with hp Finite Elements. I. One- and Two-Dimensional Elliptic and Maxwell Problems*. Chapman & Hall/CRC Press, 2006. (Cited on pp. 53, 58, 59, 60, 61, 72, 137)

[26] L. Demkowicz. Polynomial exact sequences and projection-based interpolation with applications to Maxwell equations. In D. Boffi and L. Gastaldi, editors, *Mixed Finite Elements, Compatibility Conditions and Applications*, volume 1939 of *Lecture Notes in Mathematics*, pages 101–158. Springer-Verlag, 2008. See also ICES Report 06-12. (Cited on pp. 60, 82, 97, 99)

[27] L. Demkowicz. *Lecture Notes on Energy Spaces*. Technical Report 13, ICES, 2018. (Cited on pp. xi, 4, 12, 20, 30, 42, 48, 106, 155, 183)

[28] L. Demkowicz. *Lecture Notes on Maxwell Equations in a Nutshell*. Technical Report 2, Oden Institute, 2020. (Cited on p. 28)

[29] L. Demkowicz and I. Babuška. p interpolation error estimates for edge finite elements of variable order in two dimensions. *SIAM J. Numer. Anal.*, 41(4):1195–1208, 2003. (Cited on pp. 60, 82)

[30] L. Demkowicz and A. Buffa. H^1, $H(\mathrm{curl})$ and $H(\mathrm{div})$-conforming projection-based interpolation in three dimensions. Quasi-optimal p-interpolation estimates. *Comput. Methods Appl. Mech. Engrg.*, 194:267–296, 2005. (Cited on pp. 60, 82, 97, 99)

[31] L. Demkowicz, T. Fuehrer, N. Heuer, and X. Tian. The double adaptivity paradigm (how to circumvent the discrete inf-sup conditions of Babuška and Brezzi). *Comput. Math. Appl.*, 95:41–66, August 2021. (Cited on pp. 154, 183, 190, 195)

[32] L. Demkowicz and J. Gopalakrishnan. A class of discontinuous Petrov-Galerkin methods. Part I: The transport equation. *Comput. Methods Appl. Mech. Engrg.*, 199(23–24):1558–1572, 2010. See also ICES Report 2009-12. (Cited on p. 147)

[33] L. Demkowicz and J. Gopalakrishnan. A class of discontinuous Petrov-Galerkin methods. Part II: Optimal test functions. *Numer. Methods Partial Differ. Equ.*, 27:70–105, 2011. See also ICES Report 2009-16. (Cited on p. 147)

[34] L. Demkowicz, J. Gopalakrishnan, and B. Keith. The DPG-Star method. *Comput. Math. Appl.*, 79(11):3092–3116, 2020. (Cited on p. 154)

[35] L. Demkowicz, J. Kurtz, D. Pardo, M. Paszyński, W. Rachowicz, and A. Zdunek. *Computing with hp Finite Elements. II. Frontiers: Three-Dimensional Elliptic and Maxwell Problems with Applications.* Chapman & Hall/CRC Press, 2007. (Cited on pp. 58, 59, 60, 61, 67, 72, 76, 137)

[36] L. Demkowicz, P. Monk, L. Vardapetyan, and W. Rachowicz. De Rham diagram for hp finite element spaces. *Comput. Math. Appl.*, 39(7–8):29–38, 2000. (Cited on pp. 81, 97)

[37] L. Demkowicz and J. T. Oden. Recent progress on applications of hp-adaptive BE/FE methods to elastic scattering. *Int. J. Numer. Methods Eng.*, 37:2893–2910, 1994. (Cited on p. 121)

[38] L. Demkowicz and L. Vardapetyan. Modeling of electromagnetic absorption/scattering problems using hp-adaptive finite elements. *Comput. Methods Appl. Mech. Engrg.*, 152(1–2):103–124, 1998. (Cited on pp. 32, 135)

[39] L. Demkowicz and P. Zanotti. Construction of DPG Fortin operators revisited. *Comput. Math. Appl.*, 80:2261–2271, 2020. Special Issue on Higher Order and Isogeometric Methods. (Cited on p. 172)

[40] W. Dörfler. A convergent adaptive algorithm for Poisson's equation. *SIAM J. Numer. Anal.*, 33(3):1106–1124, 1996. (Cited on p. 191)

[41] I. Ekeland and R. Temam. *Convex Analysis and Variational Problems.* North-Holland, 1976. (Cited on p. 187)

[42] I. Ergatoudis, B. M. Irons, and O. C. Zienkiewicz. Curved, isoparametric, "quadrilateral" elements for finite element analysis. *Int. J. Solids Structures*, 4:31–42, 1968. (Cited on p. 59)

[43] A. Ern and J.-L. Guermond. *Finite Elements* I. *Approximation and Interpolation*, volume 72 of *Texts in Applied Mathematics*. Springer, 2021. (Cited on p. xi)

[44] A. Ern and J.-L. Guermond. *Finite Elements* II. *Galerkin Approximation, Elliptic and Mixed PDEs*, volume 73 of *Texts in Applied Mathematics*. Springer, 2021. (Cited on p. xi)

[45] A. Ern and J.-L. Guermond. *Finite Elements* III. *First-Order and Time-Dependent PDEs*, volume 74 of *Texts in Applied Mathematics*. Springer, 2021. (Cited on p. xi)

[46] Brezzi F. and M. Fortin. *Mixed and Hybrid Finite Element Methods.* Springer-Verlag, 1991. (Cited on p. 126)

[47] F. Fuentes, B. Keith, L. Demkowicz, and S. Nagaraj. Orientation embedded high order shape functions for the exact sequence elements of all shapes. *Comput. Math. Appl.*, 70:353–458, 2015. (Cited on pp. 60, 76)

[48] I. M. Gelfand and S. V. Fomin. *Calculus of Variations.* Dover, 2000. (Cited on p. 1)

[49] J. Gopalakrishnan and L. Demkowicz. Quasioptimality of some spectral mixed methods. *J. Comput. Appl. Math.*, 167(1), 2004. (Cited on p. 103)

[50] J. Gopalakrishnan and W. Qiu. An analysis of the practical DPG method. *Math. Comp.*, 83(286):537–552, 2014. (Cited on p. 172)

[51] M. Greenberg. *Foundations of Applied Mathematics.* Prentice-Hall, 1978. (Cited on p. 2)

[52] B. M. Irons. Numerical integration applied to finite element methods. In *Conference on the Use of Digital Computers in Structural Engineering*. University of Newcastle, 1996. (Cited on p. 59)

[53] B. Keith, F. Fuentes, and L. Demkowicz. The DPG methodology applied to different variational formulations of linear elasticity. *Comput. Methods Appl. Mech. Engrg.*, 309:579–609, 2016. (Cited on p. 27)

[54] D. Kincaid and W. Cheney. *Numerical Analysis*. Brooks/Cole Publishing Company, 1996. 2nd ed. (Cited on p. 136)

[55] A. Korn. Ueber einige Ungleichungen, welche in der Theorie der elastischen und elektrischen Schwingungen eine Rolle spielen. *Bull. Internat. Acad. Sci. Cracovie*, 9:705–724, 1909. (Cited on p. 45)

[56] A. Majda. *Compressible Fluid Flow and Systems of Conservation Laws in Several Space Variables*, volume 53 of *Applied Mathematical Sciences*. Springer-Verlag, 1984. (Cited on p. 16)

[57] J. M. Melenk and C. Rojik. On commuting p-version projection-based interpolation on tetrahedra. *Math. Comp.*, 321:45–87, 2020. (Cited on pp. 82, 97)

[58] S. G. Mikhlin. *Variational Methods in Mathematical Physics*. Pergamon Press, 1964. (Cited on p. 118)

[59] P. Monk and L. Demkowicz. Discrete compactness and the approximation of Maxwell's equations in $\mathbb{R}^3$. *Math. Comp.*, 70(234):507–523, 2001. (Cited on p. 135)

[60] I. Muga and K. G. Van der Zee. Discretization of linear problems in Banach spaces: Residual minimization, nonlinear Petrov–Galerkin, and monotone mixed methods. *SIAM J. Numer. Anal.*, 58(6):3406–3426, 2020. (Cited on p. 116)

[61] S. Nagaraj, S. Petrides, and L. Demkowicz. Construction of DPG Fortin operators for second order problems. *Comput. Math. Appl.*, 74(8):1964–1980, 2017. (Cited on p. 172)

[62] J. C. Nédélec. Mixed finite elements in $\mathbb{R}^3$. *Numer. Math.*, 35:315–341, 1980. (Cited on pp. 72, 174)

[63] J. C. Nédélec. A new family of mixed finite elements in $\mathbb{R}^3$. *Numer. Math.*, 50:57–81, 1986. (Cited on p. 72)

[64] J. Nečas. *Les méthodes directes en théorie des équations elliptiques*. Masson, 1967. (Cited on p. 131)

[65] J. T. Oden, L. Demkowicz, R. Rachowicz, and T.A. Westermann. Toward a universal hp adaptive finite element strategy. Part 2: A posteriori error estimation. *Comput. Methods Appl. Mech. Engrg.*, 77:113–180, 1989. (Cited on pp. 60, 81)

[66] J. T. Oden and L. F. Demkowicz. *Applied Functional Analysis for Science and Engineering*. Chapman & Hall/CRC Press, 2018. Third edition. (Cited on pp. 3, 21, 22, 36, 114, 122, 124, 139, 149, 187)

[67] J. T. Oden and J. N. Reddy. *An Introduction to the Mathematical Theory of Finite Elements*. Dover, 2011. (Cited on p. xi)

[68] P. A. Raviart and J. M. Thomas. A mixed finite element method for 2-nd order elliptic problems. In *Mathematical Aspects of Finite Element Methods*, volume 606 of *Lecture Notes in Math.*, pages 292–315. Springer-Verlag, 1977. (Cited on p. 72)

[69] D. B. Szyld. The many proofs of an identity on the norm of oblique projections. *Numer. Algorithms*, 42:309–323, 2006. (Cited on p. 114)

[70] C. Weber. A local compactness theorem for Maxwell's equations. *Math. Methods Appl. Sci.*, 2:12–25, 1983. (Cited on p. 47)

[71] J. Xu and L. Zikatanov. Some observations on Babuška and Brezzi theories. *Numer. Math.*, 94:195–202, 2002. (Cited on pp. 115, 116)

Index